Josiah Parsons Cooke

First Principles of Chemical Philosophy

Josiah Parsons Cooke

First Principles of Chemical Philosophy

ISBN/EAN: 9783337069001

Printed in Europe, USA, Canada, Australia, Japan

Cover: Foto ©berggeist007 / pixelio.de

More available books at **www.hansebooks.com**

FIRST PRINCIPLES

OF

CHEMICAL PHILOSOPHY.

BY

JOSIAH P. COOKE, Jr.,
ERVING PROFESSOR OF CHEMISTRY AND MINERALOGY IN HARVARD COLLEGE.

CAMBRIDGE:
WELCH, BIGELOW, AND COMPANY,
PRINTERS TO THE UNIVERSITY.
1868.

Entered according to Act of Congress, in the year 1868, by
JOSIAH P. COOKE, JR.,
in the Clerk's Office of the District Court of the District of Massachusetts.

UNIVERSITY PRESS: WELCH, BIGELOW, & CO.,
CAMBRIDGE.

ERRATA.

Page 39, line —4, *for* reactions *read* reaction.
" 40, " 15, " [17] and [18] *read* [21] and [22].
" 50, " 22, " $(NH_4)_2,(Al_2)_3^2 O_8\text{-}(SO_2)_4, 24H_2O$
　　　　　　read $(NH_4)_2,(Al_2)$ viii O_8 viii $(SO_2)_4 . 24H_2O$.
" 50, " —3, *for* nitrobenzoel *read* nitrobenzole.
" 61, " —5, " $C_2H_6O_3$ " C_2H_6O.
" 63, " 24, " Oxzchloride " Oxychloride.
" 72, " 19, " forms " form.
" 73, " —8, " sextivalent " sexivalent.
" 78, " 10, " atrivalent " trivalent.
" 85, " 16, " bibasic " dibasic.
" 87, " 16, " 61 " 85.
" 92, " —11, " bibasic " dibasic.
" 95, " —6, " calcium " cæsium.
" 98, " —2, " symbols " symbol.
" 102, " 10, " 61 " 85.
" 112, " —4, " 11 " 13.
" 114, " 17, " 50 " 53.
" 121, " 4, " 56 " 59.
" 136, " 21', " bibasic " dibasic.
" 169, " 19, " 892 " § 92.
" 170, " 14, " Sp. Gr. " *Sp. Gr.*
" 170, " —10, " *R* " *r.*
" 171, " —3, " [89] " (§ 89)..
" 173, " —9, " would " could.
" 174, " 4, " 53 " 61.
" 190, " 12, " these " those.

PREFACE.

The object of the author in this book is to present the philosophy of chemistry in such a form that it can be made with profit the subject of college recitations, and furnish the teacher with the means of testing the student's faithfulness and ability. With this view the subject has been developed in a logical order, and the principles of the science are taught independently of the experimental evidence on which they rest. It is assumed that the student has already been made familiar with this evidence, and with the more elementary facts which the philosophy of the science attempts to interpret. At most of our American colleges this instruction is given in a course of experimental lectures; but for less mature students a course of manipulation in the laboratory will be found a far more efficient mode of teaching, and some preliminary training of this kind ought to be made one of the requisites for admission to our higher institutions of learning.[1]

This book is intended to supplement such a course of practical instruction. It deals solely with the theories of the science, and with those principles which can only be acquired by study and application. The author has found by long experience that a recitation on mere facts, or descriptions of apparatus and experiments, is to the great mass of college undergraduates all but worthless, while the study of the philosophy of chemistry may be made highly profitable both for instruction and discipline. Moreover, our college students

---

[1] For such a course of practical study the student can desire no better guide than the excellent work of Professors Eliot and Storer, recently published, "A Manual of Inorganic Chemistry, arranged to facilitate the Experimental Demonstration of the Facts and Principles of the Science." By C. W. Eliot and F. H. Storer. New York, 1868.

begin the study of physical science with a degree of maturity, and a kind of mental culture, which enables them to acquire that limited knowledge and general view of the subject, for which alone they have time and occasion, most rapidly when it is presented in a condensed and deductive form. The author has had especially in view this class of students, and has endeavored to meet their wants.

However important a training in the methods and the inductive logic of science may be in itself considered, it would be vain and unprofitable to attempt to change the habits of thought of those whose education has been almost wholly classical, and who are preparing themselves for a professional or literary career, where they will have occasion to use the results more than the methods of science. On the other hand, we find at our colleges a not inconsiderable portion of the students, whose tastes and abilities find their best exercise in the study of natural science, and who are preparing for the medical profession or other spheres of practical life, for which a training of the powers of observation and of inductive reasoning is an indispensable requisite. For such students the college should furnish the culture they require in a course of elective study; but beginning the study of chemistry as they do in the present organization of our colleges, at an advanced stage of their education, they will gain time if their practical work is preceded or at least accompanied by the study of what may be figuratively called the grammar of the science. Lastly, to that ever increasing class of students who seek their mental culture solely in "scientific studies," the philosophy of science is especially important; for in an exclusive devotion to facts and methods, they are not likely to gain that breadth of view and enlargement of mind which the study of theory is calculated to give. In all experimental science, theory is undoubtedly subordinate to practice, but it gives form and dignity to our knowledge, and the two should never be divorced in our systems of education.

The value of problems as means of culture and tests of attainment can hardly be overestimated, and they have therefore been made a chief feature in this book. Since those which are here given are chiefly intended as guides to the student, the answers have always been added; and where the method was not obvious, the chief steps in the solution have been given as well. Every teacher will be able to multiply problems after these models to suit his own requirements.

The questions, which accompany the problems, form another essential feature in the plan of instruction here presented. They are intended not only to direct the student's attention to the most important points, but also to stimulate thought by suggesting inferences to which the principles stated legitimately lead.

These questions, moreover, will indicate to the teacher the manner, in which it is intended that the book should be studied. Care should be taken not to overstrain the memory, but to distribute the necessary burthen through many lessons. Thus, for the first seven chapters, the student should not be expected to reproduce the symbols and reactions, nor even to call the names of the substances represented, except those of the more familiar elements and simplest compounds. It will be sufficient for the time if he understands the principles which the symbols illustrate, and the relations of the parts of the reactions, although as yet these conventional signs may have for him no more definite meaning than the paradigms of a grammar. As he advances through chapters VIII. and IX., he should be expected to familiarize himself with the names of the compounds, and should begin to reproduce the symbols. When reciting on chapter X. he should be called upon to give not only the names of all the symbols, but also the symbols corresponding to all the names, and so on for the rest of the book. In reviewing the book a full knowledge of the names and symbols will be of course expected from the first. The questions and problems appended to each chapter will give the student a clear idea of what in any case will be required. The author has been in the habit of writing out, for his own class, similar problems on separate cards, together with the names, symbols, reactions or other data, which may in any case be given. These cards are distributed at the beginning of each recitation, and the student is not called upon to recite until he has placed his work upon the blackboard. This plan obviates many practical difficulties, and has been found to work with great success.

The philosophy of chemistry has been developed in this book according to the "modern theories"; and the author would acknowledge his obligations to the recent works of Miller Frankland, Naquet, Roscoe, Williamson, and Wurtz, all of which he has freely consulted. Careful attention has been given to the chemical notation; and a method has been devised of writing rational symbols, which, while it fully exhibits the relations of the parts of the molecule, condenses the formulæ, and saves space and labor in printing. From a desire to secure uniformity, the nomenclature of the London Chemical Society has been adopted; but, in the chapter on this subject, the old names are given with the new. Lastly, the metric system of weights and measures, and the centigrade scale of the thermometer, are used throughout the book.

CAMBRIDGE, December 1, 1868.

FIRST PRINCIPLES

OF

CHEMICAL PHILOSOPHY.

PART I.

CHAPTER I.

INTRODUCTION.

1. *Definitions.* — The *volume* of a body is the space it fills, expressed in terms of an assumed unit of volume. The *weight* of a body, as the word is used in chemistry and generally in common life, is the amount of material which the body contains compared with that in some other body assumed as the unit of weight. The *specific gravity* of a body is the ratio of its weight to that of an equal volume of some substance which has been selected as the standard. Solids and liquids are always compared with water at its greatest density, which is at 4° centigrade, and hence the numbers which stand for their specific gravities express how many times heavier they are than an equal volume of water at this temperature. Gases, however, are most conveniently compared with the lightest of all known forms of matter, namely, hydrogen, and in this book the number which indicates the specific gravity of a gas expresses how many times heavier it is than an equal volume of hydrogen, compared under the same conditions of temperature and pressure.

2. *Volume and Weight.* — All experimental science rests upon accurate measurements of these fundamental elements, and it is therefore very important that there should be a general agreement among scientific men in regard to them. This

has been secured by the almost universal adoption of the French system of measures and weights in all scientific investigations. The details of this system are given in Table I., and they require no further explanation. Its great advantage over our ordinary English system is not only in its decimal subdivision, but also in the simple relation which exists between the units of measure and of weight. Since the unit of weight is the weight of the unit volume of water, and since the specific gravity of solids and liquids is always referred to water, as the standard, it is always true in this system that

$$W = V \times Sp.\ Gr. \qquad [1]$$

If the volume is given in cubic centimetres, the weight obtained is in grammes; but if the volume is given in cubic decimetres or litres, the weight is found in kilogrammes. In this formula, $Sp.\ Gr.$ stands for the specific gravity referred to water. If the specific gravity is referred to hydrogen, as in the case of gases, the value must be reduced to the water-standard before using it in the formula. The reduction is easily made, by multiplying by 0.0000896, a fraction which is simply the specific gravity of hydrogen itself referred to water. Using Sp. Gr. to represent the specific gravity of a gas referred to hydrogen, the formula becomes

$$W = V \times Sp.\ Gr. \times 0.0000896, \qquad [2]$$

and may then be used in all calculations connected with the weight and volume of aeriform bodies. In such calculations, in order to avoid the long decimal fractions which the use of the gramme entails, Hofmann has proposed to introduce into chemistry a new unit of weight which he calls the *crith*. This unit is the weight of one cubic decimetre or litre of hydrogen gas at the standard temperature and pressure, and is equal to 0.0896 grammes. If now we estimate the weight of all gases in *criths*, and let W represent this weight, while W represents the weight in grammes, and V the volume in *litres*, we shall also have

$$\mathrm{W} = \mathrm{V} \times Sp.\ Gr.\ \text{and}\ W = \mathrm{W} \times 0.0896, \qquad [3]$$

and all problems of this kind will then be reduced to their simplest terms.

The specific gravity of gases is also frequently referred to dry air, which for many reasons is a convenient standard. The weight of one litre of air under standard conditions is 1.293187 grammes. Hence, representing specific gravity referred to air by 𝔖p. 𝔊r. we have

$$\text{Sp. Gr.} : \mathfrak{Sp. Gr.} = 1.2932 : 0.0896,$$

or

$$\text{Sp. Gr.} = \mathfrak{Sp.Gr.} \times 14.42,$$

and

$$\mathfrak{Sp. Gr.} = \text{Sp. Gr.} \times 0.06929.$$

3. *Chemistry and Physics.*— Among material phenomena we may distinguish two classes. First, those which are manifested without a loss of identity in the substances involved. Secondly, those which are attended by a change of one or more of the materials employed into new substances. The science of chemistry deals with the last class of phenomena, that of physics with the first, and hence the terms chemical and physical phenomena. An illustration will make this distinction plain. When a bar of iron is drawn out into wire, is rolled out into thin leaves, is reduced by mechanical means to powder, is forged into various shapes, is melted and cast into moulds, is magnetized, or is made the medium of an electric current, since the metal does not in any case lose its identity, the phenomena are all physical. When, on the other hand, the iron bar rusts in the air, is burnt at the blacksmith's forge, or is dissolved in dilute sulphuric acid, the iron is converted into a new substance, iron rust, iron cinders, or green vitriol, and the phenomena are chemical. The distinction between these two departments of human knowledge is not, however, so strongly marked as the definition would seem to imply. In fact they coalesce at many points, and a knowledge of the elements of physics is an essential preliminary to the successful study of chemistry. In the following pages it will be assumed that the student is acquainted with the most elementary principles of this science, and references will be made to the sections of the author's work on Chemical Physics. The same relation which physics bears to chemistry on the one side, chemistry bears to physiology and the natural-history sciences on the other.

INTRODUCTION.

Questions and Problems.

1. Reduce by Table I. at the end of the book,

30 Inches to fractions of a metre.	Ans. 0.7619 metre.
76 Centimetres to inches.	Ans. 29.92 inches.
36 Kilometres to miles.	Ans. 22.38 miles.
10 Metres to feet and inches.	Ans. 32 ft. 9.7 inches.
1 Cubic metre to quarts.	Ans. 880.66 quarts.
1 Cubic foot to litres.	Ans. 28.31 litres.
1 Pint to cubic centimetres.	Ans. $567.8 \; \overline{c.\,m.}^3$
1 Litre to cubic inches.	Ans. 61.027 cubic inches.
1 Pound Avoirdupois to grammes.	Ans. 453.6 grammes.
1 Kilogramme to ounces avoirdupois.	Ans. 35.27 ounces.
1 Ounce to grammes.	Ans. 28.35 grammes.

2. If the globe were a perfect sphere what would be the circumference and what the diameter in kilometres?

Ans. Circumference 40,000 kilometres,
Diameter   12,732.4   "

3. The length of the metre was determined by measuring the distance between Dunkirk (in France), Latitude 51° 2′ 9″ and Formentera (one of the Balearic Islands), Latitude 38° 39′ 56″, both on the same meridian. This distance was found by triangulation to be equal to 730,430 toises. What is the length of a metre in terms of this old French unit of measure? What, also, was the length measured in English miles? No account is to be taken of the ellipticity of the earth.   Ans. The metre, 0.5314 toise.
The length was 854 miles.

4. The *Sp. Gr.* of iron is 7.84. What is the weight of 10 $\overline{c.\,m.}^3$ of the metal in grammes? What is also the weight in kilogrammes of a sphere of iron whose diameter equals one decimetre?

Ans. 78.4 grammes and 4.105 kilogrammes.

5. What is the weight in grammes of 50 $\overline{c.\,m.}^3$ of oil of vitriol, when the *Sp. Gr.* of the liquid is 1.8?   Ans. 90 grammes.

6. The *Sp. Gr.* of alcohol being 0.8, what volume in litres would weigh 7.2 kilogrammes?   Ans. 9 litres.

7. Assuming that the earth is spherical, and its mean *Sp. Gr.* 5.67, what would be its weight, using as the unit of weight a kilometre cube of water at its greatest density?   Ans. 6,130,000,000,000.

8. Determine the *Sp. Gr.* of absolute alcohol from the following data: — weight of empty bottle 4.326; weight of same filled with water 19.654; weight of same filled with alcohol 16.741.
Ans. 0.8095.

9. Determine the *Sp. Gr.* of lead from: — weight of empty bottle 4.326; weight of same filled with water 19.654; weight of lead shot 15.456; weight of bottle filled in part with the shot and the rest with water 33.766.   Ans. 11.5.

10. Determine the *Sp. Gr.* of iron from: — weight of iron in air 3.92; weight under water 3.42.   Ans. 7.84.

11. Determine *Sp. Gr.* of wood from: — weight of wood in air 25.35; weight of copper sinker in air 11; weight of same under water 9.77; weight of wood with sinker under water 5.10 grammes.
   Ans. 0.8445.

12. How much volume must a hollow sphere of copper have, weighing one kilogramme, which will just float in water? What must be the volume of the copper?
   Ans. One cubic decimetre and 111.8 $\overline{c.m.}^3$

13. How much volume must a hollow cylinder of iron have, which weighs 10 kilogrammes and sinks one half in water, and what must be the volume of the metal? Ans. 20 and 1.276 cubic decimetres.

14. What is the weight in grammes (under standard conditions) of 128 $\overline{c.m.}^3$ of oxygen gas (Sp. Gr. = 16)?
   Ans. 0.1834 grammes.

15. How many litres of carbonic anhydride gas (Sp. Gr. = 22) would weigh (under normal conditions) 4.480 kilogrammes?
   Ans. 2274 litres.

16. Solve the last two problems by [3], and show in what respect the method differs from that indicated by [2].

17. What is the weight in criths (under standard conditions) of one litre of nitrogen gas (Sp. Gr. = 14), of one litre of chlorine gas (Sp. Gr. = 35.5), of one litre of marsh gas (Sp. Gr. = 8), and of one litre of ammonia gas (Sp. Gr. = 8.5)?
   Ans. 14, 35.5, 8, and 8.5 criths respectively.

18. What is the weight in grammes of one litre of each of the same gases under the same conditions?
   Ans. 1.254, 3.180, 0.7165, and 0.7617 respectively.

19. The weight of one litre of hydrochloric acid gas is 1.635 grammes; of carbonic oxide gas 0.9703 grammes; of cyanogen gas 2.328 grammes, and of hydrogen gas 0.0896 grammes. What is the specific gravity of each of these gases referred to air?
   Ans. 1.265, 0.9703, 0.9007, and 0.0693 respectively.

20. What is the volume (under standard conditions) of 12.54 grammes of nitrogen gas, when specific gravity referred to air is 0.9703?   Ans. 10 litres.

21. What is the weight of one litre of air in criths?

Ans. 14.42.

22. What would be the ascensional force of one thousand litres of hydrogen, under normal conditions?

Ans. The ascensional force is the difference between the weight of the hydrogen and that of the air displaced. Hence in the present example, the ascensional force would be 14,420 — 1000 = 13420 criths, or 1,201 grammes.

23. What is the value of a crith in grains, English weight.

Ans. 1.382 grains.

CHAPTER II.

FUNDAMENTAL CHEMICAL RELATIONS.

4. *Compounds and Elements.*—With sixty-three exceptions, all known substances, by various chemical processes, may be decomposed, and hence are called *chemical compounds;* while the sixty-three substances which have as yet never been resolved into simpler parts are called *chemical elements.* There is some reason for believing that many, if not all, of these elementary substances may hereafter be decomposed, and hence they can only be considered chemical elements provisionally; but, however this may be, all known materials may still be regarded as formed by the union of the particles of one or more of these sixty-three substances. A list of the chemical elements is given in Table II. The names of the more abundant or otherwise more important elements are printed in Roman letters. The others are very rare substances, and are practically unimportant. Of these elementary substances more than three fourths possess metallic properties, and among them are all the useful metals, including the liquid metal mercury. The rest present every variety of physical character. Oxygen, hydrogen, and nitrogen are permanent gases. Chlorine, and probably fluorine, though gases under ordinary conditions, may by pressure and cold be condensed to liquids. Bromine is a very volatile liquid; and among the solids we have every gradation between the highly volatile iodine, or the easily fusible phosphorus, on the one hand, and carbon, which has never even been melted, on the other. We find, also, among the elements every difference as regards density. Hydrogen gas is the lightest, and the metal platinum the heaviest substance known. Several of the elementary substances occur in a free state in nature, for example, oxygen and nitrogen in the atmosphere, carbon in the coal beds, sulphur in the neighborhood of active volcanoes, iron in meteoric stones, while arsenic, an-

timony, bismuth, copper, gold, silver, mercury, and platinum, with a few other rare associates, are sometimes found in a more or less pure state in metallic veins. Gold and platinum are usually found in a free condition, though as a rule slightly alloyed with their associated metals; but all the other elements are generally found in combination, and the greater number appear in nature only in this condition. From such compounds the elements may be extracted by various chemical processes, which will appear as we proceed. Among these elements the useful metals are the tools of civilization, carbon is our universal fuel, while sulphur, phosphorus, arsenic, chlorine, bromine, and iodine have found important applications in the arts, and are therefore articles of commerce; but the greater number of the elements are only to be seen in the chemist's laboratory, and are solely objects of chemical investigation. The elements are distributed in nature in very unequal proportions. At least one half of the solid crust of the globe, eight ninths of the water on its surface, and one fifth of the atmosphere which surrounds it, consist of the one element, oxygen. Moreover, the other elements are usually found in combination with oxygen, so that oxygen may be regarded as the cement by which these elementary parts of the world are held together. Next in abundance is silicon, which, after oxygen, is the chief constituent of the rocks, and makes up about one fourth of the earth's crust. Silicon is always found combined with oxygen, and more than one half of the oxygen of the globe is in combination with this element. Hence, the compound of the two, which we call silica or quartz, must make up more than one half of our solid globe, at least as far as its composition is known. After silicon in the order of abundance would follow the elements aluminum, calcium, magnesium, potassium, sodium, iron, carbon, sulphur, hydrogen, chlorine, nitrogen, which, without attempting to discriminate between them, make up altogether very nearly the other fourth of the earth's mass; for the remaining fifty elements — including all the useful metals except iron — do not constitute altogether more than one one-hundredth. Of the sixty-three known elements, then, thirteen alone make up at least $\frac{9.9}{100}$ of the whole known mass of the earth.

5. *Analysis and Synthesis.* — The composition of a chemical

compound may be made evident in two ways. First, by breaking up the compound into its constituent parts; secondly, by reuniting these parts and reproducing the original substance. The first of these methods of proof is called *analysis*, the second, *synthesis*. The study of the processes by which the composition of a body may be discovered, and the relative amounts of its various constituents determined, forms an important branch of practical chemistry, which is known as *Chemical Analysis*, and this is subdivided into Qualitative and Quantitative Analysis, according to the object we have in view. Synthesis is chiefly used to prove the results of analysis.

6. *Law of Definite Proportions.* — Numberless analyses have proved that *any given chemical compound always contains the same elements combined in the same proportions*. Thus, when we analyze water, sugar, and salt, we always obtain the result given below; and this result is invariable, saving small errors of observation, from whatever source these materials may be drawn. The composition is given in per cents, as is usual in such cases.

Water (Dumas).	Salt.	Sugar (Péligot).
Hydrogen, 11.112	Sodium, 39.32	Carbon, 42.06
Oxygen, 88.888	Chlorine, 60.68	Hydrogen, 6.50
		Oxygen, 51.44
100.	100.	100.

Chemists have not yet succeeded in making sugar by combining its elements, but the synthesis both of water and salt is easily effected, and illustrates still more, forcibly the same law. Thus we may *mix* together hydrogen and oxygen gas in *any* proportion, but when, by passing an electric spark through the mixture, we cause the elements to combine, then the gases unite in the exact proportion indicated above, and any excess of one or the other which may be present is left over. The law of definite proportions gives to chemistry a mathematical basis; for, since the analyses of all compounds have been made and tabulated in a way that will be soon explained, it is always possible, when the weight of a compound is given, to calculate the weights of its constituents, and, when the weight of one of its elements is known, to calculate the weights of all the other elements present.

7. *Mixture and Chemical Compound.* — The law of definite proportions gives a simple criterion for distinguishing between a mixture and a true chemical compound. In the first the elements may be mixed in any proportion, but in the true compound they are always combined in definite proportions. Thus we may mix together copper-filings and sulphur in any proportion, but as soon as we apply heat, and cause the elements to combine, then the copper combines with one half of its own weight of sulphur, and the excess of either element above these proportions is discarded. Again, in a mixture however homogeneous, we can generally, by mechanical means alone, distinguish the ingredients. Thus, in the mixture just referred to, a microscope would show the grains of sulphur and metallic copper, with all their characteristic appearances; and by means of carbonic sulphide we can easily dissolve out all the sulphur from the mixture; but after the chemical union has taken place, the characteristic properties of the elements are *merged* in those of the compound, and no such simple mechanical separation is possible. But although these distinctions are generally sufficient, nevertheless we find in some alloys, in solutions, and in a few other classes of compounds, less intimate conditions of chemical union where these criterions fail.

8. *Law of Multiple Proportions.* — It is generally the case that the same elements unite in more than one proportion, forming two or more different compounds. Now we always find that the proportions of the elements in such compounds are simple multiples of each other. This law is best illustrated by the compounds of nitrogen and oxygen, which are five in number, and have the names indicated in the table below. In order to make evident the law, we give, not the percentage composition as above, but the amount of oxygen, which is in each case combined with one and three fourths parts of nitrogen.

COMPOUNDS OF NITROGEN WITH OXYGEN.

	Nitrogen. By weight.	Oxygen. By weight.	Nitrogen. By volume.	Oxygen. By volume.
Nitrous Oxide,	1.75	1	2	1
Nitric Oxide,	1.75	2	2	2
Nitrous Anhydride,	1.75	3	2	3
Nitric Peroxide,	1.75	4	2	4
Nitric Anhydride,	1.75	5	2	5

CHAPTER III.

MOLECULES.

9. *Molecules.* — In order to bring the facts of chemistry into relation with each other, and unite them in an harmonious system, the following theory, first proposed by the English chemist, Dalton, and known as the Atomic Theory, is generally accepted by chemists. This theory assumes, in the first place, that every body, whatever its substance may be, is formed by the aggregation of minute particles of the same kind, which cannot be further subdivided without destroying the *identity* of the substance. Thus a lump of sugar is an aggregate of minute particles of sugar. If the sugar is burnt, these particles will be further subdivided; but the sugar will be thus changed into new substances. In like manner, a drop of water is an aggregate of minute particles of water. By passing a current of electricity through the drop, these particles will be subdivided, but then we shall have no longer water, but the two elementary gases, oxygen and hydrogen. *The smallest particles of any substance which can exist by themselves, we call molecules.*

10. *Physical Properties of Matter.* — The physical qualities of a body depend solely on the relations of its molecules. The physicist has therefore no occasion to continue the subdivision beyond the molecule, which is his unit.

Solid. — In a solid the molecules firmly cohere, and the force which binds them together has been called cohesion. On the form and size of the molecules, and also on the mode of aggregation, is supposed to depend the crystalline form of each substance, which is one of the most important and characteristic properties of matter, and one to which we shall have occasion hereafter to refer. On certain relations of the molecules, which we do not fully understand, depend undoubtedly elasticity, tenacity, ductility or malleability, hardness, transparency, diathermancy, and the allied qualities of solid bodies.

Liquid. — In the liquid condition of matter the molecules have more freedom of motion than in the solid, but still the motion is circumscribed within the liquid mass. Moreover, a certain cohesion still exists between the molecules, and on this depends the form of the rain-drop. The various phenomena of capillary action also are effects of the *cohesion* of the liquid molecules modified by their *adhesion* to the surfaces of solids, and the solvent power of liquids is a still further effect of the same mutual action. Connected also with this freedom of molecular motion is the property of liquids of transmitting pressure in all directions, and the well-known principles of hydrostatics to which it leads; but this property belongs to the third condition of matter as well.

Gas. — In the aeriform condition of matter, the motion of the molecules is only circumscribed by the walls of the containing vessel, or by some force acting on the mass from without. The molecules of a gas are constantly beating against the walls which confine them, and were they not thus restrained would fly off into space. The molecules of the atmosphere are restrained by the force of gravitation, and, as they fly upwards like a ball thrown into the air, they are at last brought to rest, and fall back again to the earth. Hence gases always exert pressure against any surface with which they are in contact, and we measure the pressure, or, as we frequently call it, the tension of the gas, by the height to which it will raise a column of mercury. Chem. Phys. (158). The instrument used for this purpose is called a barometer.

The height of the mercury column which represents the pressure or tension of a gas is always represented by H.

In our latitude, at the surface of the sea, the atmosphere in its normal conditions will raise a column of mercury 76 c. m. high. Hence $H = 76$, and to this standard we always refer in comparing together the volumes of different gases.

11. *Mariotte's Law.* — The most characteristic feature of the aeriform condition is the great change of volume which gases undergo, under varying pressure, and the special law of compressibility which they obey. If we represent by H and H' two conditions of pressure to which the *same body* of gas is at different times exposed, then the law is expressed by the formula

$$V : V' = H' : H. \qquad [4]$$

Moreover, since the specific gravity of a *given mass* of gas must be the greater the less its volume, it is also true that

$$Sp.\ Gr. : Sp.\ Gr'. = H : H', \qquad [5]$$

and lastly, since the weight of a *given volume* of gas is obviously proportional to its specific gravity, we also have

$$W : W' = H : H', \qquad [6]$$

in which W and W' represent the weight of an equal volume of the same gas under the two pressures H and H'.

12. *Heat a Manifestation of Molecular Motion.* — The effects of what we call heat are supposed to be merely manifestations of the motion of the molecules of bodies. The greater the moving power of the molecule, the more forcibly it strikes against our nerves of feeling, and hence the more intense is the sensation of heat produced; and to the condition of matter which produces this sensation we give the name of *temperature*. The greater the moving power of the molecules, the higher the temperature; the less the moving power, the lower the temperature. Moreover, since by the very definition all molecules at the *same temperature* are in the condition to produce the *same sensation* of heat, we must assume further, that, whatever their size or weight, they must all have, at the same temperature, the same moving power. The light molecule of hydrogen must move much faster than the heavy molecule of carbonic anhydride in order to produce the same effect. If now we represent the mass of any molecule by m, and by V its velocity at any given temperature, then the moving power will be represented by $\tfrac{1}{2}m\ V^2$, Chem. Phys. (42), and this will have the same value for every molecule at the same temperature. With a few exceptions, all bodies expand with an increasing temperature, and in the case of mercury the change of volume is so nearly proportional to the change of temperature that we may use the varying volume of a confined mass of this liquid as a measure of temperature. This is the simple theory of the common mercurial thermometer, and in this book we shall refer all temperatures to the degrees of the centigrade scale. These degrees are purely arbitrary; but to each one corresponds a definite value of $\tfrac{1}{2}m\ V^2$, although we have not as yet been able to connect our arbitrary with our theoretical measure.

When we increase the temperature of a body, we must of course increase the moving power of all the molecules, each by the same amount, and the sum of the moving powers which they thus acquire is the legitimate measure of the amount of heat which the body receives. Hence, while $\frac{1}{2}m\ V^2$ represents the temperature of a body, $\Sigma\ \frac{1}{2}m\ V^2$ represents the whole amount of heat which it contains. Practically, however, we measure quantity of heat by an arbitrary standard, and we shall use in this book as our unit the amount of heat required to raise the temperature of a kilogramme of pure water from 0° to 1° centigrade. This we call the *Unit of Heat*, and it has been found, by careful experiments, that this unit of heat represents an amount of moving power which is adequate to raise a weight of 423 kilogrammes one metre, or to do any other equivalent amount of work.

13. *Expansion by Heat.* — The amount of expansion which bodies undergo when heated has been carefully measured for many different substances, and the results are tabulated in all works on physics. Chem. Phys. Table XV. In each case is given the coefficient of expansion, which is the small fraction of its volume which a body increases when heated one centigrade degree. If, now, K represents this fraction, V the initial volume, V' the new volume, t the initial temperature, and t' the new temperature, then, if we assume that the expansion is proportional to the temperature, we easily deduce the formula, Chem. Phys. (239),

$$V' = V(1 + K(t' - t)). \qquad [7]$$

This formula serves to calculate the change of volume both of solids and gases, which expand, nearly at least, proportionally to the temperature. The same, however, is not true of liquids, whose rate of expansion frequently increases, with the temperature, very rapidly; and for such bodies we are obliged to use the following formula, which is of the general form in which every algebraic function may be developed, and is much less simple: —

$$V' = V(1 + At + Bt^2 + Ct^3 + \&c.). \qquad [8]$$

In this formula, V' represents the required volume at some temperature, t, and V, the volume at 0°, which is assumed to be known; while A, B, C, &c., are numerical constants, which have been determined by experiment in the case of most liquids. Chem. Phys. (255).

Both solids and liquids expand with irresistible force, and we have, therefore, only this one effect to consider in regard to the action of heat upon them. It is different, however, with gases. By enclosing a gas in a tight vessel, we can raise its temperature without changing its volume, except so far as the vessel itself becomes enlarged by the heat. The effect of the heat is, then, to increase the tension or pressure of the gas. Hence, in the case of a gas, we may have two distinct effects; first, an increase of volume, when the pressure is constant; secondly, an increase of tension, when the volume is constant. The increased volume may always be calculated from the initial volume and difference of temperature, by means of the formula,

$$V' = V(1 + 0.00366\,(t' - t)), \qquad [9]$$

which differs from that just given only in that the numerical value has been substituted for K, — this being the same for all gases. On the other hand, the increased tension may always be calculated from the initial tension, by means of the corresponding formula,

$$H' = H(1 + 0.00366\,(t' - t)), \qquad [10]$$

in which H and H' stand for the heights of the mercury columns which measure the initial and final tension respectively. The last formula is easily deduced from the first, on the principles of Mariotte's law, stated above. Chem. Phys. (261) and [201].

Variations of temperature produce such important changes in the volume and specific gravity of all bodies, and especially of gases, that it becomes frequently essential, before comparing together different observations, to reduce them all to some standard temperature. Most scientific men use, as this standard temperature, 0° centigrade, and scientific measures are generally adjusted to this standard; but 60° Fahrenheit, corresponding to 15°.5 centigrade, is often a more convenient standard, because it is nearer the mean temperature of the air, and is, therefore, not unfrequently employed.

14. *Change of State.* — Many substances are capable of ex-

isting in all the three conditions of matter. Now, we find that whenever a solid changes to a liquid, or a liquid to a gas, heat is absorbed; and conversely, whenever a gas is liquefied, or a liquid becomes a solid, heat is evolved; although, as a general rule, this change of state is accompanied by no change of temperature. Thus, one kilogramme of ice, in melting, absorbs 79 units of heat, although the temperature remains at 0° during the change; and when, by boiling, a kilogramme of water is converted into steam, under the normal pressure of the air, no less than 537 units of heat disappear, although the temperature both of the steam and of the water is constant at 100° during the whole period. On the other hand, when the steam is condensed or the water frozen, absolutely the same amount of heat is set free as was before absorbed. The heat thus absorbed or set free is generally called the *latent heat* of the liquid or gas, and in the case of many substances the amount has been carefully measured. Chem. Phys. (277) and (299). According to the theory we are studying, these effects are the direct results of the molecular condition of matter. The change of state must be accompanied by a change in the relative position of the molecules, or in their distance from each other; and this change, in its turn, must be attended with a destruction or production of the moving power on which the effects of heat depend. Chem. Phys. (215 bis.).

15. *Sources of Heat.* — The sun is the original source of almost all the heat we enjoy on the earth, for the effect of the earth's internal heat, at its surface, is at best very small, — and all our artificial sources of heat have drawn their supply either directly or indirectly from the great central luminary. According to our theory the effect of the sun's rays is a simple result of the transfer of molecular motion from the sun to the earth, either by some unknown influence exerted from a distance, or else by an actual transfer of motion through the material particles of the ether, which is assumed to fill the intervening space. The great source of all artificial heat is combustion in its many forms, and this, as we shall hereafter see, is merely a clashing together of material molecules, and is necessarily attended with a great development of that moving power to which we refer all thermal effects.

16. *Specific Heat.* — The amount of heat required to raise

to the same extent the temperature of equal weights of different substances is by no means the same. The quantity is capable in any case of exact measurement, and is called the specific heat of the substance. The amount of heat required to raise the temperature of one kilogramme of water one centigrade degree has been assumed as the unit, and we express the specific heat of other substances in terms of this measure. Moreover, since with the exception of hydrogen the specific heat of water is greater than that of any substance known, the specific heat of all other bodies must be expressed by fractional numbers. In every case, unless otherwise stated, the numbers indicate what fraction of a unit of heat would be required to raise the temperature of one kilogramme of the substance from $0°$ to $1°$ centigrade. Chem. Phys. (232).

17. *Molecular Condition of Gases.* — The aeriform state is by far the simplest condition of matter, and there are two peculiarities in its properties which lead to important conclusions in regard to its molecular conditions. These characteristics are as follows: First, All true gases obey the same law of compressibility. Secondly, Equal volumes of all true gases expand equally on the same increase of temperature. Chem. Phys. (262). Now according to our theory these peculiar relations of the aeriform condition of matter can only be explained on the assumption that *Equal volumes of all gases contain the same number of molecules.* It can easily be seen that the properties just enumerated would be a necessary consequence of this fact, and this important theoretical deduction lies at the basis of our modern theories of Chemistry. This peculiar molecular condition, however, is only found in the gas, for it is only in this state that the molecules are sufficiently separated from each other to be freed from the mutual action of those molecular forces which give rise to far more complicated relations in both liquid and solid bodies. Moreover, with our ordinary gases (in the degree of condensation in which they exist under the pressure of the atmosphere), the molecules are not yet sufficiently far apart to be wholly freed from the effects of their mutual action, and hence the theoretical condition is not absolutely fulfilled; and in vapors, where the molecules are still closer together, the variation from the theory is quite large. In proportion as the gas ex-

pands, the theoretical condition is approached, and, when in a state of great expansion, equal volumes of all gases would undoubtedly contain exactly the same number of molecules. It is only then that we reach the condition of what we have called above the *true gas*, and this is our criterion of its state, — that it obeys absolutely the law of Mariotte. A very important corollary follows at once from the principle we have just deduced.

The molecular weight of all substances is directly proportioned to their specific gravities in the state of gas.

We have adopted in this book hydrogen gas as our unit of specific gravity for aeriform substances, and were we also to take the molecule of hydrogen as our unit of molecular weight, then the number which expresses the specific gravity of a gas would express also its molecular weight. But for reasons which will appear hereafter, we have selected the half hydrogen molecule as our unit, and hence *the molecular weight of any substance in terms of this unit is always twice its specific gravity in the state of gas.* In Table III. we have given, according to the most accurate experimental data, the Sp. Gr. (referred to hydrogen) of all the best known gases and vapors, and in a parallel column we have also given the Half-molecular Weights of the same substances determined by chemical analysis, in a manner which will be hereafter described. It will be seen that the numbers in the second column are almost precisely the same as those in the first, and the slight differences which will be noticed, either arise from the fact that the vapors, under the conditions in which alone their Sp. Gr. can be accurately determined, are not true gases, that is, do not exactly obey Mariotte's law; or in other cases, where the differences are more considerable, may be referred to a partial decomposition of the substance itself in the process of the experiment. In solving the problems of this book, and generally in most chemical problems, the Half-molecular weight may be taken as the true Sp. Gr. The logarithms of these values given in the last column of the table will be found useful in this connection. Although only given to four places of decimals, they exceed in accuracy the experimental data. The values in the column of Sp. Gr. referred to air, are given, as a rule, to one decimal place beyond the limit of error.

Questions and Problems.

1. Are the qualities of a molecule of any substance, the same as those which distinguish the substance itself?

2. What is the distinction between cohesion and adhesion?

3. When the barometer stands at 76 c. m., with what weight in grammes is the air pressing against each square centimetre of surface? *Sp. Gr.* of mercury 13.596. Ans. 1033.

4. To what difference of pressure does a difference of one centimetre in the barometric column correspond?
Ans. 13.596 grammes.

5. When a mercury barometer stands at 76 c. m. how high would a water barometer stand? Also, how high would barometers stand filled with alcohol or sulphuric acid, disregarding in each case the tension of the vapor? *Sp. Gr.* of alcohol 0.81; *Sp. Gr.* of sulphuric acid 1.85. Ans. 1033; 1275 and 558.2 c. m.

6. A volume of hydrogen gas was found to be 200 $\overline{c. m.}^3$ The height of the barometer observed at the same time, was 74 c. m. What would have been the volume if observed when the barometer stood at 76 c. m. Ans. 194.7 $\overline{c. m.}^3$

7. A volume of nitrogen standing in a bell-glass over a mercury pneumatic trough measured 250 $\overline{c. m.}^3$ The barometer at the time stood at 75.4 c. m., and the level of the mercury in the bell was found by measurement to be 6.5 above the surface of the mercury in the trough. Required to reduce the volume to standard pressure.

Ans. The pressure of the air on the surface of the mercury in the trough (measured at 75.4 c. m.) was balanced first by the column of mercury in the bell, and secondly by the tension of the confined gas. Hence the pressure to which the gas was exposed was equal to $75.4 - 6.5 = 68.9$ c. m. and we have $76 : 68.9 = 250 : x = 226.7$ $\overline{c. m.}^3$

8. What would be the answer to the same problem, had the trough been filled with water?

Ans. The water column in the bell exerts a pressure which is as much less than the pressure of the mercury column in the previous problem, as the *Sp. Gr.* of water is less than the *Sp. Gr.* of mercury. Hence we have $13.6 : 1 = 6.5 : 0.48$, also $75.4 - 0.48 = 74.92$, and $76 : 74.92 = 250 : x = 246.4$ $\overline{c. m.}^3$

9. A closed vessel, which displaces one litre of air, is poised on a balance with weights, whose volume is inconsiderable when compared with that of the vessel. The balance is in equilibrium when

the barometer stands at 76 c. m. If the barometer falls to 73 c. m. how much weight must be added to restore the equilibrium?
Ans. 85 milligrammes.

10. Given the weight of one litre of dry air under the normal conditions as 14.42 criths, what will be the weight of one litre of dry air at the normal temperature, but under a pressure of 72 c. m.?
Ans. 13.67 criths.

11. A volume of gas measures 500 $\overline{\text{c. m.}}^3$ at 15° what will be its volume at 288°.2? In this and the next three problems the pressure is assumed to be constant.
Ans. 1000 $\overline{\text{c. m.}}^3$

12. To what temperature must an open vessel be heated before one quarter of the air which it contains at 0° is driven out?
Ans. 91°.07.

13. An open vessel is heated to 819°.6. What portion of the air which the vessel contained at 0° remains in it at this temperature?
Ans. ¼.

14. A closed glass vessel, which at 13° was filled with air having a tension of 76 c. m. is heated to 559°.4. Determine the tension of the heated air.
Ans. 3 atmospheres.

15. Reduce the following volumes of gas measured at the temperatures and pressure annexed to 0° and 76 c. m.

1. 210° $\overline{\text{c. m.}}^3$ $H = 57$ c. m. $t = 136°.6$ Ans. 70 $\overline{\text{c. m.}}^3$

2. 320° $\overline{\text{c. m.}}^3$ $H = 95$ c. m. $t = 91°.1$ Ans. 192 $\overline{\text{c. m.}}^3$

3. 480° $\overline{\text{c. m.}}^3$ $H = 38$ c. m. $t = 68°.3$ Ans. 96 $\overline{\text{c. m.}}^3$

16. What is the weight of dry air contained in a glass globe of 640 $\overline{\text{c. m.}}^3$ capacity at the temperature 546°.4 and under a pressure of 71.25 c. m.
Ans. 0.2583 grammes.

General Solution. — In order to make the solution general we will represent the capacity of the globe, the temperature and the height of the barometer by V, t and H respectively. We can also easily find from Table III. that one cubic centimetre of dry air at 0°, and when the barometer stands at 76 c. m., weighs 14.42 criths or 0.001292 grammes. To find what one cubic centimetre would weigh when the barometer stands at H centimetres, we make use of proportion [6], whence we derive

$$w = 0.001292 \cdot \frac{H}{76},$$

the weight of one cubic centimetre at 0° and under a pressure of H centimetres. To find what one cubic centimetre would weigh under the same pressure but at $t°$, it must be remembered that one cubic centimetre at $0°$ becomes $(1 + t\,0.00366)$ cubic centimetres at $t°$ [7]; therefore at $t°$ and at H centimetres of the barometer $(1 + t\,0.00366)$ $\overline{\text{c. m.}}^3$ weigh $0.00129 \cdot \dfrac{H}{76}$ grammes. By equating these two terms we obtain

$$(1 + t\,0.00366) = 0.00129 \cdot \frac{H}{76},$$

whence

$$1 = 0.00129 \cdot \frac{1}{1 + t\,0.00366} \cdot \frac{H}{76}.$$

the weight of one cubic centimetre at $t°$ and under a pressure of H centimetres. The weight of V cubic centimetres (w) is evidently

$$w = 0.00129\ \text{V} \cdot \frac{1}{1 + t\,0.00366} \cdot \frac{H}{76}. \qquad [10\,a]$$

Thus far in this solution we have neglected the change in capacity of the glass globe due to the change of temperature. This causes no sensible error when the change of temperature is small, but when the change of temperature is quite large the change of capacity of the globe must be considered. If the capacity is V $\overline{\text{c. m.}}^3$ at $0°$ it becomes at $t°$ .V $(1 + t\,0.00003)$. (See Chem. Phys. §§ 241 – 244.) Introducing this value for V into the above equations we obtain

$$w = 0.00129\ \text{V}\,(1 + t\,0.00003) \cdot \frac{1}{1 + t\,0.00366} \cdot \frac{H}{76}. \qquad [10\,b]$$

17. Required a general method for determining the Sp. Gr. of a vapor.

Solution. — The specific gravity of a vapor has been defined as its weight compared with the weight of the same volume of hydrogen gas under the same conditions of temperature and pressure, but practically it is most convenient to determine the Sp. Gr. with reference to air, and subsequently to reduce the result to the hydrogen standard.

To find, then, the Sp. Gr. of a vapor, we must ascertain the weight of a known volume, V, at a known temperature, t, and under a known pressure, H, and divide this by the weight of the same volume of air at the same temperature, and under the same pressure. The method may best be explained by an example. Suppose, then, that we wish to ascertain the Sp. Gr. of alcohol vapor. We take a light glass globe having a capacity of from 400 to 500 $\overline{\text{c. m.}}^3$, and draw the neck out in the flame of a blast lamp, so as to leave only a fine opening, as shown in the figure at a. The first step is now to ascertain the weight of the glass globe when completely exhausted of air. As this cannot readily be done directly, we weigh the globe full of air, and then subtract the weight of the air, ascertained by calculation from the capacity of the globe, and from the temperature and pressure of the air, by means of equation (10 a). Call the weight of the globe and air W, and the weight of the air w, then W—w is the weight of the globe exhausted of air. The second step is to ascertain the weight of the globe filled with alcohol vapor at a known temperature, and under a known pressure. For this purpose we introduce into the globe a few grammes of pure alcohol, and mount it on the support represented in the accompanying figure. By loosening the screw, r, we next sink the balloon beneath the oil contained in the iron vessel, V, and secure it in this position. We now slowly raise the temperature of the oil to between 300° and 400°, which we observe by means of the thermometer, T. The alcohol changes to vapor, and drives out the air, which, with the excess of vapor, escapes at a. When the bath has attained the requisite temperature, we close the opening a, by suddenly melting the end of the tube at a with a mouth blowpipe, and as nearly as possible at the same moment observe the temperature of the bath and the height of the barometer. We have now the globe filled with alcohol vapor at a known temperature, and under a known pressure. Since it is hermetically sealed, its weight cannot change, and we can therefore allow it to cool, clean it, and weigh it at our leisure. This will give us the weight of the globe filled with alcohol vapor at a known temperature, t', and under a known pressure, H'. Call this weight W'. The weight of the vapor is W' — W + w. The third step is to ascertain the weight of the same volume of air at the same temperature and under the same pressure. This can easily be found by calculation from equation (10 b). The last step is to find the capacity of the globe, which, although we have supposed it known, is not actually ascertained experimentally until the end of the process. For this purpose we break off the tip of the tube (a), under mercury, which, if the experiment has been carefully conducted, rushes in and fills the globe completely. We then empty this mercury into a carefully graduated glass cylinder, and read off the volume. We find then the

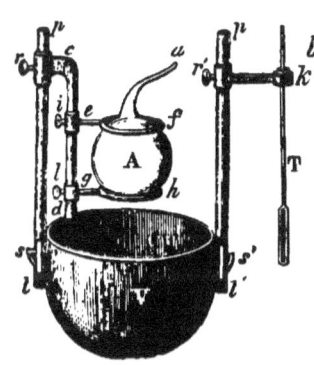

MOLECULES. 23

Sp. Gr. by dividing the weight of the vapor by the weight of the air. The formulæ for the calculation are then

Weight of the globe and air,   W.

" " air, $w = 0.001292 \text{ V} \cdot \dfrac{1}{1 + t\, 0.00366} \cdot \dfrac{\text{H}}{76}$.

" " globe exhausted of air,   W — w.

" " " filled with vapor at a temperature t' and under a pressure H',   W'.

" " vapor,   W' — W + w.

" " air at t' and under a pressure H', =
$0.001292 \text{ V} (1 + t'\, 0.00003) \cdot \dfrac{1}{1 + t'\, 0.00366} \cdot \dfrac{\text{H}'}{76}$.

$$\text{Sp.Gr.} = \dfrac{\text{W}' - \text{W} + w}{0.001292 \text{ V} (1 + t'\, 0.00003) \cdot \dfrac{1}{1 + t'\, 0.00366} \cdot \dfrac{\text{H}'}{76}}$$

1. Ascertain the Sp. Gr. of alcohol vapor from the following data: —

Weight of glass globe,	W	50.804 grammes.
Height of barometer,	H	74.75 centimetres.
Temperature,	t	18°
Weight of globe and vapor,	W'	50.824 grammes.
Height of barometer,	H'	74.76 centimetres.
Temperature,	t'	167°
Volume,	V	351.5 cubic centimetres.

Ans. 1.575.

2. Ascertain the Sp. Gr. of camphor vapor from the following data: —

Weight of glass globe,	W	50.134 grammes.
Height of barometer,	H	74.2 centimetres.
Temperature,	t	13°.5
Weight of globe and vapor,	W'	50.842 grammes.
Height of barometer,	H'	74.2 centimetres.
Temperature,	t'	244°
Volume,	V	295 cubic centimetres.

Ans. 5.371.

CHAPTER IV.

ATOMS.

18. *Definition.* — The atomic theory assumes that so long as the identity of a substance is preserved its molecules remain undivided; but when, by some chemical change, its identity is lost, and new substances are formed, the theory supposes that the molecules themselves are broken up into still smaller particles, which it calls *atoms*. Indeed it regards this division of the molecules as the very essence of a chemical change.

The word atom is derived from α, privative, and τεμνω (I cut), and recalls a famous controversy in regard to the infinite divisibility of matter, which for many centuries divided the philosophers of the world. But chemistry does not deal with this metaphysical question. It asserts nothing in regard to the possible divisibility of matter; but its modern theories claim that, practically, this division cannot be carried beyond a certain extent, and that we then reach particles which cannot be further divided by any chemical process now known. These are the chemical atoms, and the atom is simply the unit of the chemist, just as the molecule is the unit of the physicist, or the stars the units of the astronomer. The molecule is a group of atoms, and is a unit in the microcosm, of which it is a part, in the same sense that the solar system is a unit in the great stellar universe. The molecule has been defined as the smallest particle of any substance which can exist by itself, and the atom may be now defined as *the smallest mass of an element that exists in any molecule*.

When a molecule breaks up, it is not supposed that the atoms fall apart like grains of sand; but simply that they arrange themselves in new groups, and thus give rise to the formation of new substances. Indeed, as a rule, the atoms cannot exist in a free state, and with few exceptions every molecule consists of at least two atoms. This is thought to be true, even of the chemical elements. The difference between the molecules of an elementary substance and those of a compound, according to the theory, is merely this, that while the first are formed by the union of atoms of the same kind, the last comprise atoms of different kinds. The molecules of oxygen gas are atomic aggregates as well as those of water, only the molecules of oxygen consist of oxygen atoms alone, while the molecules of water contain both oxygen and hydrogen atoms. Such at least is the constitution of most elementary substances. Nevertheless, in the case of mercury, zinc, cadmium, and some other metallic elements, the facts compel us to believe that the molecule consists of but one atom, or, in other words, that in these cases the molecule and the atom are the same.

19. *Atomic Weights.* — There must be evidently as many kinds of atoms as there are elementary substances; and, since these substances always unite in definite proportions, it must be also true that the elementary atoms have definite weights. This once assumed, the law of multiple proportions, as well as that of definite proportions, becomes an essential part of our atomic theory; for, since the atoms are by definition indivisible, the elements can only combine atom by atom, and must therefore unite either in the proportion of the atomic weights or in some simple multiples of this proportion. We have discovered no means of measuring even approximately the absolute weight of an atom; but, after we have determined, from considerations hereafter to be discussed, what must be the number of atoms of each kind in one molecule of any substance, we can easily calculate their relative weight from the results of analysis. A few examples will make the method plain.

1. The analysis of water, given on page 6, proves that in 100 parts it contains 11.112 parts of hydrogen and 88.888 parts of oxygen. Every molecule of water, then, must contain these two elements in just these proportions. Now we have good reason for believing that each molecule of water is a group of three atoms, — two of hydrogen and one of oxygen. Then, since $\frac{1}{2}$ (11.112) : 88.888 $=$ 1 : 16, it follows that the oxygen atom must weigh 16 times as much as the hydrogen atom; and, if we make the hydrogen atom the unit of our atomic weight, then the weight of the oxygen atom, estimated in these units, must be 16.

2. The analysis of hydrochloric acid gas proves that it con-

tains in 100 parts 2.74 parts of hydrogen and 97.26 of chlorine. Moreover, we have reason to believe that each molecule of the acid is a group of two atoms, — one of hydrogen and one of chlorine. Hence the atom of chlorine must weigh 35.5 times as much as that of hydrogen. Its atomic weight is then 35.5.

3. The analysis of common salt, page 6, proves that it contains in 100 parts 60.68 parts of chlorine and 39.32 parts of sodium, and we believe that each molecule of salt is a group of two atoms, one of chlorine and one of sodium. Then, since $60.68 : 39.32 = 35.5 : 23$, it follows that the atomic weight of sodium is 23. In like manner the atomic weights of all the chemical elements have been determined, and the numbers are given in Table II. These numbers are the fundamental data of chemical science, and the basis of almost all the numerical calculations which the chemist has to make. The elements of a compound body are always united either in the proportions, by weight, expressed by these numbers, or else in some simple multiples of these proportions; and whenever, by the breaking up of a complex compound, or by the mutual action of different substances on each other, the elements rearrange themselves, and new compounds are formed, the same numerical proportions are always preserved.

The atomic weights evidently rest on two distinct kinds of data; *first*, on the results of chemical analysis, which are facts of observation, and in regard to which the only question can be as to their greater or less accuracy; *secondly*, on our conclusions in regard to the number of atoms in each molecule of the substance analyzed. This conclusion again is based chiefly on two classes of facts, whose bearing on the subject we must briefly consider.

1. In the first place we carefully compare together all the compounds of the element we are studying, with the view of discovering the smallest weight of it which enters into the composition of any known molecule; for this must evidently be the atomic weight of the element. An example will make the course of reasoning intelligible.

In the following table we have a list of a number of the most important compounds containing hydrogen, all of which either are gases, or can easily be changed into vapor by heat, so that their specific gravities in the state of gas can be readily determined. From these specific gravities we learn the weights of the molecules (compare § 17) which are given in the second column of the table. In the third column we have given the weight of hydrogen contained in the molecules, referred, of course, to the same unit as the weight of the molecules themselves: —

Compounds of Hydrogen.	Weight of Molecule referred to Hydrogen Atom.	Weight of Hydrogen in the Molecule.
Hydrochloric Acid	36.5	1
Hydrobromic Acid	81.0	1
Hydriodic Acid	128.0	1
Hydrocyanic Acid	27.0	1
Hydrogen Gas	2.0	2
Water	18.0	2
Sulphuretted Hydrogen	34.0	2
Seleniuretted Hydrogen	81.5	2
Formic Acid	46.0	2
Ammonia	17.0	3
Phosphuretted Hydrogen	34.0	3
Arseniuretted Hydrogen	78.0	3
Acetic Acid	60.0	4
Olefiant Gas	28.0	4
Marsh Gas	16.0	4
Alcohol	46.0	6
Ether	74.0	10

Assuming now, as has been assumed in this table, that a molecule of hydrogen gas weighs 2, it appears that the smallest mass of hydrogen which the molecule of any known substance contains, weighs just one half as much, or 1. We infer, therefore, that this mass of hydrogen cannot be divided by any chemical means, or, in other words, that it is the hydrogen atom. The molecule of hydrogen gas contains then two hydrogen atoms, and this atom is the unit to which we refer all molecular and atomic weights.

If now, in like manner, we bring into comparison all the volatile compounds of oxygen, we shall find that the smallest mass of oxygen which exists in the molecule of any known substance weighs 16, — the atom of hydrogen weighing 1, — and hence we infer that this mass of oxygen is the oxygen atom. Moreover it will appear that a molecule of oxygen gas weighs

32, and hence it follows that each molecule of oxygen gas, like the molecule of hydrogen, is formed by the union of two atoms. A similar comparison would show that, while the molecule of nitrogen gas weighs 28, the atom weighs 14, so that here again the molecule consists of two atoms. This method of investigation can be extended to a large number of the chemical elements, and the conclusions to which it leads are evidently legitimate, and cannot be set aside, until it can be shown that some substance exists whose molecule contains a smaller mass of any element than that hitherto assumed as the atomic weight, or, in other words, until the old atom has been divided.

2. The second class of facts on which we rely for determining the number of atoms in a given molecule is based on the specific heat of the elements (compare § 16). It would appear that the specific heat is the same for all atoms, and, if this is true, we might expect that equal amounts of heat would raise to the same extent the temperatures of such quantities of the various elementary substances as contain the same number of atoms, provided, of course, that these atomic aggregates are compared under the same conditions. Now we can determine accurately the number of units of heat required to raise the temperature of equal weights of the elementary substances one degree, and the results, which we call the specific heat of the elements, are given in works on physics. Chem. Phys. (232). Evidently, if our principle is true, these values must be proportional in every case to the number of atoms of each element contained in the equal weights compared. Representing then by S and S' the specific heat of two elementary substances, by m and m' the weights of the corresponding atoms, and by unity the equal weights compared, we shall have, in any case,

$$S : S' = \frac{1}{m} : \frac{1}{m'}, \text{ or } mS = m'S', \qquad [11]$$

that is, *The product of the atomic weight of an elementary substance by its specific heat is always a constant quantity.*

Taking now the atomic weights obtained by the method first given, and the specific heats of the elements as they have been determined by experimenting on these substances in the solid correct; and this principle not only frequently enables us to fix the atomic weight of an element, when the first method fails, but it also serves to corroborate the general accuracy of our results. It is true, owing undoubtedly to many causes which influence the thermal conditions of a solid body, that this product is not absolutely constant. It varies between 5.7 and 6.9, the mean value being about 6.34 (see Table IV). But the variation is not important, so far as the determination of the atomic weights is concerned. This determination, as we have seen, rests chiefly on the results of analysis. The question always is only between two or three possible hypotheses, and as between these the specific heat will decide. For example, an analysis of chloride of silver proves that each molecule contains for one atom, or 35.5 parts of chlorine, 108 parts of silver. Now, 108 parts of silver may represent one, two, three, or four atoms, or it may be that this quantity only represents a fraction of an atom. To determine, we divide 6.34 by 0.057, the specific heat of silver. The result is 111, which, though not the exact atomic weight, is near enough to show that 108 is the weight of one atom, and not of two or three. The exceptions to this rule referred to above are carbon, boron, and silicon. But the specific heat of these elements varies so very greatly with the differences of physical condition — the so-called allotropic modifications — which these elements present, — Chem. Phys. (234), — that the exceptions are not regarded as invalidating the general principle. The law simply fails in these cases, and we can see why it fails.

This important law, whose bearing on our subject we have briefly considered, was first discovered by Dulong and Petit, and was subsequently verified by the very careful experiments of Regnault. More recently it has been found, by Voestyn and others, that its application extends, in some cases at least, to chemical compounds; for it would seem that the atoms retain, even when in combination, their peculiar relations to heat, so that the product of the specific heat of a substance by its molecular weight is equal to as many times 6.3 as there are atoms in the molecule. Thus the specific heat of common salt, multiplied by its molecular weight, gives $0.214 \times 58.5 = 12.52$, which is very nearly equal to 6.3×2; while in the case of corrosive sublimate the corresponding product, $0.069 \times 271 = 18.70$, is nearly equal to 6.3×3, — results which are in accordance with our views in regard to the number of atoms in the molecules of these substances.

We have here, then, an obvious method by which we might determine the number of atoms in the molecule of any solid, and which would be of the very greatest value in investigating the atomic weights, could we rely on the general application of our law. We do not expect mathematical exactness. We know very well that the specific heat of solid bodies varies very greatly with the temperature, as well as from other physical causes, and that it is impossible to compare them under *precisely the same* conditions, as would be required in order to secure accuracy. But, unfortunately, the discrepancies are so great, and we are so ignorant of their cause, that as yet we have not been able to place much reliance on the specific heat as a means of determining the number of atoms in the molecules of a compound.

3. Lastly, assuming that both of the means we have considered fail to give satisfactory evidence in regard to the number of atoms in the molecule of a given substance (which we may have analyzed for the purpose of determining some atomic weight), we may frequently, nevertheless, reach a satisfactory, or at least a probable conclusion, by comparing the substance we are investigating with some closely allied substance whose constitution is known. Thus, if the molecule of sodic chloride (common salt) contains two atoms, it is probable that the molecules of sodic iodide, as well as those of potassic chloride and potassic iodide, contain the same number; for all these compounds not only have the same crystalline form and the same chemical relations, but also they are composed of closely allied chemical elements. Nevertheless it is true, in very many cases, that our conclusion in regard to the number of atoms which a molecule may contain is more or less hypothetical, and hence liable to error and subject to change. This uncertainty, moreover, must extend to the atomic weights of the elements, so far as they rest on such hypothetical conclusions.

If we change the hypothesis in any case, we shall obtain a different atomic weight; but then the new weight will be some simple multiple of the old, and will not alter the important relations to which we first referred. These fundamental relations are independent of all hypothesis, and rest on well-established laws.

The atomic weights are the numerical constants of chemistry, and in determining their value it is necessary to take that care which their importance demands. The essential part of the investigation is the accurate analysis of some compound of the element whose atomic weight is sought. The compound selected for the purpose must fulfil several conditions. It must be one which can be prepared in a condition of absolute purity. It must be one the proportions of whose constituents can be determined with the greatest accuracy by the known methods of analytical chemistry. It must contain a second element whose atomic weight is well established. Finally, it should be a compound whose molecular condition is known, and it is best that this should be as simple as possible. When they are once thus accurately determined, the atomic weights become essential data in all quantitative analytical investigations.

Questions and Problems.

1. Does the integrity of a substance reside in its molecules or in its atoms?

2. We find by analysis that in 100 parts of potassic chloride there are 52.42 parts of potassium and 47.58 parts of chlorine. Moreover, we know from previous experiments that the atomic weight of chlorine is 35.5, and we have reason to believe that every molecule of the compound consists of two atoms, one of potassium and one of chlorine. What is the atomic weight of potassium? Ans. 39.1.

3. We find by analysis that in 100 parts of phosphoric anhydride there are 43.66 parts of phosphorus and 56.34 parts of oxygen. Moreover, we know that the atomic weight of oxygen is 16; and we have reason to believe that every molecule of the compound consists of seven atoms, 2 of phosphorus and 5 of oxygen. What is the atomic weight of phosphorus? Ans. 31.

4. In Table III. the student will find the molecular weights of the following oxygen compounds; and we give below, following the name, the weight of oxygen (estimated like the molecular weight in hydrogen atoms) which each contains. From these data it is

required to determine the atomic weight of oxygen. Oxygen Gas, 32; Water, 16; Sulphurous Anhydride, 32, Sulphuric Anhydride, 48; Phosphoric Oxychloride, 16; Carbonic Oxide, 16; Carbonic Anhydride, 32; Osmic Anhydride, 64; Nitrous Oxide, 16; Nitric Oxide, 16; and Nitric Peroxide, 32. Ans. 16.

5. We give below the weight of chlorine in one molecule of several of its most characteristic volatile compounds. It is required to deduce the atomic weight of chlorine on the principle of the last problem. Chlorine gas, 71; Phosphorous Chloride, 106.5; Phosphoric Oxychloride, 106.5; Arsenious Chloride, 106.5; Phosgene Gas, 71; Stannic Chloride, 142; Stanno-triethylic Chloride, 35.5; and Hydrochloric Acid, 35.5. Ans. 35.5.

6. Review the steps of the reasoning by which the atomic weights have been deduced in the last two problems, and show that the "*molecular weight*" and "*the weight of the element in one molecule*" are actual and independent experimental data.

7. Analysis shows that in 100 parts of mercuric chloride there are 73.80 parts of mercury and 26.20 parts of chlorine. The specific heat of mercury is 0.032. What is the probable atomic weight of mercury, that of chlorine being 35.5? Also, how many atoms of each element does one molecule of the compound contain?

Ans. Atomic weight of mercury, 200. Each molecule consists of one atom of mercury and two of chlorine.

8. Analysis shows that in 100 parts of ferric oxide there are 70 parts of iron and 30 parts of oxygen. The specific heat of iron is 0.514. What is the probable atomic weight of iron, that of oxygen being 16? and also, how many atoms of each element does one molecule of the oxide contain?

Ans. Atomic weight of iron, 56. One molecule of ferric oxide contains 2 atoms of iron and 3 of oxygen.

9. The molecular weight of silicic chloride is 170, and its specific heat, 0.1907. How many atoms does one molecule of the compound probably contain? Ans. 5.

10. The molecular weight of mercuric iodide is 454, and its specific heat, 0.042. How many atoms does one molecule of the compound probably contain? Ans. 3.

CHAPTER V.

CHEMICAL NOTATION.

20. *Chemical Symbols*. — The atomic theory has found expression in chemistry in a remarkable system of notation, which has been of the greatest value in the study of the science. In this system, the initial letter of the Latin name of an element is used as the symbol of that element, and represents in every case *one atom*. Thus O stands for *one atom* of Oxygen, N for *one atom* of Nitrogen, H for *one atom* of Hydrogen. When several names have the same initial, we add for the sake of distinction a second letter. Thus C stands for one atom of Carbon, Cl for one atom of Chlorine, Ca for one atom of Calcium, Cu for one atom of Cuprum (copper), Cr for one atom of Chromium, Co for one atom of Cobalt, Cd for one atom of Cadmium, Cs for one atom of Cæsium, and Ce for one atom of Cerium. The symbols of all the elements are given in Table II. Several atoms of the same element are generally indicated by adding figures, but distinguishing them from algebraic exponents by placing them below the letters. Thus Sn_2 stands for two atoms of Stannum (tin), S_3 for three atoms of Sulphur, and I_5 for five atoms of Iodine. Sometimes, however, in order to indicate certain relations, we repeat the symbol with or without a dash between them, thus $H\text{-}H$ represents a group of two atoms of Hydrogen, $Se\text{=}Se$ a group of two atoms of Selenium. We can now easily express the constitution of the molecule of any substance by simply grouping together the symbols of the atoms of which the molecule consists. This group is generally called the symbol of the substance, and stands in every case for one molecule. Thus $NaCl$ is the symbol of common salt, and represents one molecule of salt. H_2O is the symbol of water, and represents, as before, one molecule. So in like manner H_3N stands for one molecule of ammonia gas, H_4C for one molecule of marsh gas, KNO_3 for one molecule of saltpetre, H_2SO_4 for one molecule of sulphuric acid,

$C_2H_4O_2$ for one molecule of acetic acid, $H\text{-}H$ for one molecule of hydrogen gas. We do not, however, always write the symbols in a linear form, but group the letters in such a way as will best indicate the relations we are studying. When several molecules of the same substance take part in a chemical change, we represent the fact by writing a numerical coefficient before the molecular symbol. A figure so placed always multiplies the whole symbol. Thus $4H\text{-}NO_3$ stands for four molecules of nitric acid, $3\,C_2H_6O$ for three molecules of alcohol, $6\,O\text{=}O$ for six molecules of oxygen gas. When clearness requires it, we enclose the symbol of the molecule in parentheses, thus, $4(H_3\text{≡}N)$, or $(H_3\text{≡}N)_4$. The precise meaning of the dashes will hereafter appear. They are used, like punctuation marks, to point off the parts of a molecular symbol, between which we wish to distinguish.

21. *Chemical Reactions.* — These chemical symbols give at once a simple means of representing all chemical changes. As these changes almost invariably result from the *reaction* of one substance on another, they are called *Chemical Reactions.* Such reactions must necessarily take place between molecules, and simply consist in the breaking up of the molecules and the rearrangement of the atoms in new groups. In every chemical reaction we must distinguish between the substances which are involved in the change and those which are produced by it. The first will be termed the factors and the last the products of the reaction. As matter is indestructible, it follows that *The sum of the weights of the products of any reaction must always be equal to the sum of the weights of the factors*, and, further, that *The number of atoms of each element in the products must be the same as the number of atoms of the same kind in the factors.* This statement seems at first sight to be contradicted by experience, since wood and many other combustibles are consumed by burning. In all such cases, however, the apparent annihilation of the substance arises from the fact that the products of the change are invisible gases; and, when these are collected, their weight is found to be equal, not only to that of the substance, but also, in addition, to the weight of the oxygen from the air consumed in the process. As the products and factors of every chemical change must be equal, it follows that *A chemical reaction may always be represented in an equation*

CHEMICAL NOTATION. 35

by writing the symbols of the factors in the first member and those of the products in the second. Thus, the following equation expresses the reaction of dilute sulphuric acid on zinc, by which hydrogen gas is commonly prepared. The products are a solution of zinc sulphate and hydrogen gas.

$$\mathbf{Zn} + (H_2SO_4 + Aq) = (ZnSO_4 + Aq) + \mathbb{H\text{-}H}. \quad [12]$$

The initial letters of the Latin word Aqua are here used simply to indicate that the substances enclosed with it in parentheses are in solution. The symbol $\mathbf{Zn}$ is printed in "full-faced" type to indicate that the metal is used in the reaction in its well-known solid condition; while the symbol of the molecule of hydrogen is printed in skeleton type to indicate the condition of gas. This usage will be followed throughout the book; but, generally, when it is not important to indicate the condition of the materials involved in the reaction, ordinary type will be used. The molecule of hydrogen gas consists of two atoms, as our reaction indicates, and this is the smallest quantity of hydrogen which can either enter into or be formed by a chemical change. The molecule of zinc is known to consist of only one atom. When the molecular constitution of an element is not known, we simply write the atomic symbol in the reaction.

Among chemical reactions we may distinguish at least three classes. First, Analytical Reactions, in which a complex molecule is broken up into simpler ones. Thus, when sodic bisulphate is heated, it breaks up into sodic sulphate and sulphuric anhydride, —

$$Na_2S_2O_7 = Na_2SO_4 + SO_3. \quad [13]$$

So, also, by fermentation grape sugar or glucose breaks up into alcohol and carbonic anhydride, —

$$C_6H_{12}O_6 = 2C_2H_6O + 2CO_2. \quad [14]$$

Secondly, Synthetical Reactions, in which two molecules unite to form a more complex group. Thus baryta burns in an atmosphere of sulphuric anhydride, and forms baric sulphate, —

$$BaO + SO_3 = BaSO_4. \quad [15]$$

In like manner ammonia enters into direct union with hydrochloric acid to form ammonic chloride, —

$$H_3N + HCl = H_4NCl. \quad [16]$$

Thirdly, Metathetical Reactions, in which the atoms of one molecule change place with the dissimilar atoms of another, one atom of one molecule replacing one, two, three, or more atoms of the other, as the case may be. Thus, when we add a solution of common salt to a solution of argentic nitrate, we obtain a white precipitate [1] of argentic chloride, while sodic nitrate remains in solution. The result is obtained by a simple interchange between an atom of silver and an atom of sodium, as the following reaction shows: —

$$(NaCl + AgNO_3 + Aq) = (NaNO_3 + Aq) + \mathbf{AgCl}. \quad [17]$$

In the next example, one atom of barium changes place with two atoms of hydrogen. Baric chloride and sulphuric acid yield hydrochloric acid and insoluble baric sulphate, which is precipitated from the solution in water as the reaction indicates, —

$$(BaCl_2 + H_2SO_4 + Aq) = (2HCl + Aq) + \mathbf{BaSO_4}. \quad [18]$$

Of the three classes of chemical reactions the last is by far the most common, and many chemical changes which were formerly supposed to be examples of simple analysis or synthesis are now known to be the results of metathesis. In very many cases, however, a chemical reaction cannot be explained in either of these ways alone, but seems to consist in a primary union of two or more molecules and a subsequent splitting up of this large group. Indeed, this is the best way of conceiving of all metathetical reactions, for we do not suppose that in any case there is an actual transfer of atoms from one molecule to the other. The word metathesis is merely used to indicate the result of the process, not the manner in which the change takes place, and the same is true of the words analysis and synthesis.

[1] The separation of a solid or sometimes of a liquid substance in a fluid menstruum, as the result of a chemical reaction, is called precipitation, and the material which separates, a precipitate; and this, too, even when the material, being lighter than the fluid, rises instead of falls.

The common method of preparing carbonic anhydride is to pour a solution of hydrochloric acid on small lumps of marble (calcic carbonate),—

$$CaCO_3 + (2HCl + Aq) = (CaCO_3\ H_2Cl_2 + \quad [19]$$
$$Aq) = (CaCl_2 + H_2O + Aq) + CO_2.$$

We may suppose that the molecules of the two substances are, in the first place, drawn together by the force which manifests itself in the phenomena of adhesion,[1] but that, as they approach, a mutual attraction between their respective atoms comes into play, which, the moment the molecules come into collision, causes the atoms to arrange themselves in new groups. The groups which then result are determined by many causes whose action can seldom be fully traced; but there are two conditions which, *when the substances are in solution*, have a very important influence on the result. These conditions may be thus stated:—

1. Whenever a compound can be formed, which is insoluble in the menstruum present, this compound always separates as a precipitate.

2. Whenever a gas can be formed, or any substance which is volatile at the temperature at which the experiment is made, this volatile product is set free.

The reactions 17 and 18 of this section are examples of the first, while the reactions 12 and 19 are examples of the second of these conditions. The facts just stated illustrate an important truth, which must be carefully borne in mind in the study of chemistry. A chemical equation differs essentially from an algebraic expression. Any inference which can be legitimately drawn from an algebraic equation must, in some sense, be true. It is not so, however, with chemical symbols. These are simply expressions of observed facts, and, although important inferences may sometimes be drawn from the mere form of the expression, yet they are of no value whatever unless confirmed by experiment. Moreover, the facts

---

[1] We find it convenient to distinguish between the force which holds together different molecules and that which unites the atoms of the molecules. To the last we give the name of chemical affinity, while we call the first cohesion or adhesion, according as it is exerted between molecules of the same kind or those of a different kind.

which are expressed in this peculiar system of notation are as purely materials for the memory as if they were described in common language.

22. *Compound Radicals.* — In many chemical reactions the elementary atoms change places, not with other elementary atoms, but with groups of atoms, which appear to sustain relations to the compounds they leave or enter similar to those of the elements themselves. Thus, if we add to a solution of argentic nitrate a solution of ammonic chloride, we get the reaction expressed by the equation

$$AgNO_3 + NH_4\text{-}Cl = NH_4\text{-}NO_3 + AgCl. \quad [20]$$

Here the group NH_4 has taken the place of Ag. So, also, in the reaction of hydrochloric acid on common alcohol, the group C_2H_5 in the molecule of alcohol changes places with the atom of hydrogen in the molecule of hydrochloric acid, —

$$\underset{\text{Alcohol.}}{C_2H_5\text{-}O\text{-}H} + HCl = H\text{-}O\text{-}H + \underset{\text{Ethylic Chloride.}}{C_2H_5\text{-}Cl.} \quad [21]$$

We write the symbols in this peculiar way in order to make it evident to the eye that such a substitution has taken place. Lastly, in the reaction of chloroform on ammonia, the group CH of the first changes places with the three atoms of hydrogen of ammonia gas, —

$$\underset{\text{Chloroform.}}{CH\text{:}Cl_3} + H_3N = 3HCl + \underset{\text{Hydrocyanic Acid.}}{CH\text{:}N.} \quad [22]$$

Such groups as these are called compound radicals. Like the atoms themselves, they cannot, as a rule, exist in a free state; but aggregates of these radicals may exist, which sustain the same relation to the radicals that elementary substances hold to the atoms. Thus, as we have a gas chlorine consisting of molecules, represented by $Cl\text{-}Cl$, so there is a gas cyanogen consisting of molecules, represented by $CN\text{-}CN$, where CN is a compound radical called cyanogen. Again, the important radicals CO, SO_2, and PCl_3, are also the molecules of well-known gases. These *radical substances* correspond to the elementary substances previously mentioned, in which the molecule is a single atom.

But with few exceptions the radical substances have never

CHEMICAL NOTATION.

been isolated, and the radicals are only known as groups of atoms which pass and repass in a number of chemical reactions. Indeed, in the same compound we may frequently assume several radicals. The possible radicals of a chemical symbol correspond in fact almost precisely to the possible factors of an algebraic formula, and in writing the symbol we take out the one or the other, as the chemical change we are studying requires. A number of these radicals have received names, and among those recognized in mineral compounds a few of the most important are

Hydroxyl	HO	Sulphuryl	SO_2
Hydrosulphuryl	HS	Carbonyl	CO
Ammonium	H_4N	Phosphoryl	PO
Amidogen	H_2N	Nitrosyl	NO
Cyanogen	CN	Nitryl	NO_2

The radicals recognized in organic compounds are very numerous, and will be tabulated hereafter.

Questions and Problems.

1. For what do the following symbols stand?

$N; \quad Ca_2; \quad H\text{-}H; \quad H_4C; \quad 4HNO_3; \quad (C_2H_4O_2)_3.$

2. For what do the following symbols stand?

$Cl; \quad S_3; \quad O\text{=}O; \quad H_3N; \quad H_2SO_4; \quad 3C_2H_6O.$

3. For what do the following symbols stand?

$O; \quad H_5; \quad Se\text{=}Se; \quad NaCl; \quad H_2O; \quad 3KNO_3.$

4. Analyze the following reaction. Show that the same number of atoms are represented on each side of the equation, and state the class to which it belongs.

$$\mathbf{Fe} + \underset{\text{Hydrochloric Acid}}{(2HCl + Aq)} = \underset{\text{Ferrous Chloride.}}{(FeCl_2 + Aq)} + \mathrm{H\text{-}H}.$$

5. Analyze the following reaction. Show in what the equality consists, and state the class to which the reaction belongs.

$$\underset{\text{Ammonic Nitrate.}}{N_2H_4O_3} = \underset{\text{Water.}}{2H_2O} + \underset{\text{Nitrous Oxide.}}{N_2O}.$$

6. Analyze the following reactions. Show in what the equality consists, and state the class to which the reaction belongs.

$$\underset{\text{Carbon.}}{C} + \underset{\text{Oxygen.}}{O\text{=}O} = \underset{\text{Carbonic Anhydride.}}{CO_2}.$$

7. Analyze the following reaction. Show in what the equality consists, and state the class to which the reaction belongs.

$$2H\text{-}O\text{-}H \underset{\text{Water.}}{} + Na\text{-}Na \underset{\text{Sodium.}}{} = 2Na\text{-}O\text{-}H \underset{\text{Sodic Hydrate.}}{} + H\text{-}H.$$

8. The following reaction may be so written as to indicate that the products are formed by a metathesis between two similar molecules. It is required to show that this is possible.

$$2H_3N \underset{\text{Ammonia gas.}}{} = 3H\text{-}H \underset{\text{Hydrogen gas.}}{} + N\text{-}N. \underset{\text{Nitrogen gas.}}{}$$

9. Write the reactions [17] and [18] so as to indicate the manner in which the metathesis is supposed to take place.

10. State the conditions which determine the metathesis in the various reactions given in this chapter so far as these conditions are indicated.

11. Write the reactions [17] and [18] so as to indicate the manner in which the metathesis is supposed to take place.

12. Analyze the following reaction. Show what determines the metathesis and also what is meant by a compound radical.

$$(Pb\text{=}(NO_3)_2 \underset{\text{Plumbic Nitrate.}}{} + 2NH_4\text{-}Cl \underset{\text{Ammonic Chloride.}}{} + Aq) =$$

$$\mathbf{PbCl_2} \underset{\text{Plumbic Chloride.}}{} + (2NH_4\text{-}NO_3 \underset{\text{Ammonic Nitrate.}}{} + Aq)$$

13. Compare with [22] the following reaction and point out the two radicals, which, as we may assume, hydrocyanic acid contains.

$$(Ag\text{-}NO_3 \underset{\text{Argentic Nitrate.}}{} + H\text{-}CN \underset{\text{Hydrocyanic Acid.}}{} + Aq) = \mathbf{Ag\text{-}CN} \underset{\text{Argentic Cyanide.}}{} + (H\text{-}NO_3 \underset{\text{Nitric Acid.}}{} + Aq)$$

14. When sulphuric anhydride (SO_3) is added to water (H_2O) a violent action ensues and sulphuric acid is formed. The reaction may be written in two ways, and it is required to explain the different views of the process, which the following equations express.

$$H_2O + SO_3 = H_2SO_4$$
or $$2H\text{-}O\text{-}H + SO_2\text{=}O = H_2\text{=}O_2\text{=}SO_2 + H_2\text{=}O.$$

15. State the distinction between a chemical element and an elementary substance. Give also the distinction between a compound radical and a radical substance.

16. Give the names of the following radicals.

$$HO; \ HS; \ NH_4; \ NH_2; \ SO_2; \ CO; \ PO; \ NO_2, \ \&c.$$

CHAPTER VI.

STOCHIOMETRY.

23. *Stochiometry*. — The chemical symbols enable us not only to represent chemical changes, but also to calculate exactly the amounts of the substances required in any given process as well as the amounts of the products which it will yield. Each symbol stands for a definite weight of the element it represents, that is, for the weight of an atom; but, as only the relative values of these weights are known, they are best expressed as so many parts. Thus H stands for 1 part by weight of hydrogen, the unit of our system. In like manner O stands for 16 parts by weight of oxygen, N for 14 parts by weight of nitrogen, C for 12 parts by weight of carbon, C_5 for 60 parts by weight of carbon, and so on for all the symbols in Table II. The weight of the molecule of any substance must evidently be the sum of the weights of its atoms, and is easily found, when the symbol is given, by simply adding together the weights which the atomic symbols represent. Thus H_2O stands for $2 + 16 = 18$ parts of water, H_3N for $3 + 14 = 17$ parts of ammonia gas, and $C_2H_4O_2$ for $24 + 4 + 32 = 60$ parts of acetic acid.[1]

Having then given the symbol of a substance, it is very easy to calculate its percentage composition. Thus, as in 60 parts of acetic acid there are 24 parts of carbon, in 100 parts of the acid there must be 40 parts of carbon, and so for each of the other elements. The result appears below; and in the same way the percentage composition both of alcohol and ether has been calculated from the accompanying symbol.

[1] In this book "*the molecular weight of a substance*" will always mean the sum of the atomic weight of the atoms composing *one* molecule, and we shall use the phrase, "*the molecular weight of a symbol*," or "*the total atomic weight of a symbol*," to denote the sum of the atomic weights of all the molecules which the symbol represents.

	Acetic Acid $C_2H_4O_2$.	Alcohol C_2H_6O.	Ether $C_4H_{10}O$.
Carbon	40.00	52.18	64.86
Hydrogen	6.67	13.04	13.52
Oxygen	53.33	34.78	21.62
	100.00	100.00	100.00

The rule, easily deduced, is this: *As the weight of the molecule is to the weight of each element, so is one hundred parts to the percentage required.*

On the other hand, having given the percentage composition, it is easy to calculate the number of atoms of each element in the molecule of the substance. This problem is evidently the reverse of the last, but it does not, like that, always admit of a definite solution; for, while there is but one percentage composition corresponding to a given symbol, there may be an infinite number of symbols corresponding to a given percentage composition. For example, the percentage composition of acetic acid corresponds not only to the formula $C_2H_4O_2$, given above, but also to any multiple of that formula, as can easily be seen by calculating the percentage composition of CH_2O, $C_3H_6O_3$, $C_4H_8O_4$, &c. They will all necessarily give the same result, and, before we can determine the absolute number of atoms of each element present, we must have given another condition, namely, the sum of the weights of the atoms, or, in other words, the molecular weight of the substance. When this is known, the problem can at once be definitely solved.

Suppose we have given the percentage composition of alcohol, as above, and also the further fact that its molecular weight is 46. We can then at once make the proportion

$100 : 52.18 = 46 : x = 24$ the weight of the atoms of carbon,
$100 : 13.04 = 46 : x = 6$ "   "   "   "   "   " hydrogen,
$100 : 34.78 = 46 : x = 16$ "   "   "   "   "   " oxygen.

Then it follows that

$\frac{24}{12} = 2$ the number of atoms of carbon in one molecule,
$\frac{6}{1} = 6$ "   "   "   "   hydrogen in one molecule,
$\frac{16}{16} = 1$ "   "   "   "   oxygen in one molecule.

It is evident from this example, that, in order to determine exactly the symbol of a compound, we must know its molecular weight. When the substance is a gas, or is capable of being changed into vapor, we can easily determine its molecular weight by the principle on page 15. The molecular weight is simply twice its specific gravity referred to hydrogen. For all the problems given in this book, which deal only with the common gases and vapors, the molecular weight can be at once taken from Table III. If we are dealing with a new substance, we must determine its specific gravity experimentally by one of the methods which will hereafter be described.

When, on account of the fixed nature of the substance, the last mode of investigation is impossible, we can still frequently determine with great probability the molecular weight, by studying the chemical reactions into which the substance enters, and connecting, by careful quantitative experiments, the molecular weight sought with that of some substance whose molecular weight is known. The methods used in such cases will be indicated hereafter; but even when all such means fail, we can nevertheless always find which of all possible symbols expresses the composition of the substance we are studying in the simplest terms, in other words, with the fewest number of atoms in the molecule. Suppose the substance to be cane sugar, which cannot be volatilized without decomposition, and of which no reaction is known which gives any definite clew to its molecular weight. Péligot's analysis, cited on page 6, shows that it contains, in 100 parts, 42.06 parts of carbon, 6.50 parts of hydrogen, and 51.44 parts of oxygen. Assume for the moment that the molecular weight is equal to 100 then

$$\frac{42.06}{12} = 3.50 \text{ the number of atoms of carbon.}$$

$$\frac{6.50}{1} = 6.50 \text{ " " " " " hydrogen.}$$

$$\frac{51.44}{16} = 3.22 \text{ " " " " " oxygen.}$$

This would be the number of atoms of each element if the sum of the atomic weight, that is, the molecular weight, of sugar, were equal to 100. As, from the very definition, fractional atoms cannot exist, these numbers are impossible, but any other possible number of atoms must be either a multiple or a submultiple of the numbers found; and we can easily discover the fewest number of whole atoms possible, by seeking for the three smallest whole numbers which stand to each other in the relation of 3.50 : 6.50 : 3.22, a proportion which is very nearly satisfied by 12 : 22 : 11. Hence, the simplest possible symbol is $C_{12}H_{22}O_{11}$, and this has been adopted by chemists as the symbol of cane sugar, although, from anything we as yet know, the symbol may be a multiple of this. If now, taking this symbol as our starting-point, we calculate the percentage composition which would exactly correspond to it, we obtain the following results, which we have arranged in a tabular form, so that the student may compare the theoretical composition with the numbers Péligot obtained by actual analysis.

Composition of Cane Sugar,
$C_{12}H_{22}O_{11}$.

	Péligot's Analysis.	Theoretical.
Carbon	42.06	42.11
Hydrogen	6.50	6.43
Oxygen	51.44	51.46
	100.00	100.00

The difference between the two is now seen to be within the probable errors of analysis, and this example illustrates the method of arranging analytical results generally adopted by chemists.

From the above discussion we can easily deduce a simple arithmetical rule for finding the symbol of a compound when its percentage composition is known. But this rule may be best expressed in an algebraic formula, which will show to the eye at once the relation of the quantities involved in the calculation, and enable us to extend our method to the solution of many classes of problems which we might not otherwise foresee. Let us then represent

By M the weight of any chemical compound in grammes.
" m the molecular weight of the compound in hydrogen atoms.
" W the weight of any constituent of that compound, whether element or compound radical, in grammes.
" w the total atomic weight of element or radical in one molecule.

Then

$\dfrac{w}{m}$ = proportion by weight of the constituent in the compound,

and

$M\dfrac{w}{m}$ = weight of constituent in M grammes of compound, or

$$W = M\dfrac{w}{m}.\qquad [23]$$

Any three of these quantities being given, the fourth can, of course, be found. Thus we may solve four classes of problems.

1. We may find the weight of any constituent in a given weight of a compound, when we know the molecular weight of the compound, and the total atomic weight of the constituent in one molecule.

Problem. It is required to find the weight of sulphuric anhydride SO_3 in 4 grammes of plumbic sulphate PbO, SO_3. Here, $w = 32 + 3 \times 16 = 80$, $m = 207 + 16 + 80 = 303$, and $M = 4$. Ans. 1.056 grammes.

2. We can find the weight of a compound which can be produced from, or corresponds to, a given weight of one of its constituents, when the same quantities are known as above.

Problem. How many grammes of crystallized green vitriol, $FeSO_4 \cdot 7H_2O$, can be made from 5 grammes of iron? Here, $w = 56$, $m = 278$, $W = 5$. Ans. 24.821.

3. We can find the molecular weight of a compound when we have given the weight of one constituent in a given weight of the compound, and the total atomic weight of that constituent in the molecule.

Problem. In 7.5 grammes of ethylic iodide, there are 6.106 grammes of iodine; the total atomic weight of iodine in one molecule is 127. What is the molecular weight of ethylic iodide? Ans. 156.

4. We can find the total atomic weight of one constituent of a molecule when the molecular weight is given, and also the weight of the constituent in a known weight of the compound.

46   STOCHIOMETRY.

Problem. The molecular weight of acetic acid is 60, the per cent of carbon in the compound 40. What is the total atomic weight of carbon in one molecule? Ans. 24. Whence number of carbon atoms in one molecule, 2.

The last problem is essentially the same as that of finding the symbol of a compound when its percentage composition is given, while the first corresponds to the reverse problem of deducing the percentage composition from the symbol. By a slight change the formula can be much better adapted to this class of cases. For this purpose we may put $M = 100$, since we are solely dealing with per cents, and also put $w = na$, a standing for the atomic weight of any element, and n for the number of atoms of that element in one molecule of the compound we are studying. We then have

$$W = 100 \frac{na}{m} \text{ and } n = \frac{W}{100} \frac{m}{a}. \quad [24]$$

The first of these forms is adapted for calculating the per cent of each element of a compound when the molecular weight, the number of atoms of each element in one molecule, and the several atomic weights, are known; and it is evident that all these data are given by the chemical symbol of the compound. The second of these forms enables us to calculate the number of atoms of each element present in one molecule of a compound when the percentage composition, the molecular weight, and the several atomic weights, are known, and illustrates the principle before developed, that the molecular weight is an essential element of the problem.

24. *Stochiometrical Problems.* — The principles of the previous section apply not only to single molecular formulæ, but obviously may also be extended to the equations which represent chemical changes. Since the molecular symbols which are equated in these expressions represent known relative weights, it must be true in every case that we can calculate the weight of either of the factors or products of the chemical change it represents, provided only that the weight of some one is known. If we represent by w and m the total atomic weight of any two symbols entering into the chemical equations, and by W and M the weight in grammes of the factors or products

which these symbols represent, then the simple algebraic formulæ of the last section will apply to all stochiometrical problems of this kind, as well as to those before indicated. These formulæ, however, are merely the algebraic expression of the familiar rule of three, and all stochiometrical problems are solved more easily by this simple arithmetical rule. Using the word symbol to express the sum of the atomic weights it represents, we may state the rule as applied to chemical problems in the following words, which should be committed to memory.

Express the reaction in the form of an equation; make then the proportion, As the symbol of the substance given is to the symbol of the substance required, so is the weight of the substance given to x, the weight of the substance required; reduce the symbols to numbers, and calculate the value of x.

This rule applies equally well to all problems, like those of the last section, in which the elements or radicals of the same molecular symbol are alone involved; only in such cases there is of course no equation to be written. A few examples will illustrate the application of the rule.

Problem 1. We have given 10 kilogrammes of common salt, and it is required to calculate how much hydrochloric acid gas can be obtained from it by treating with sulphuric acid. The reaction is expressed by the equation

$$(2NaCl + H_2SO_4 + Aq) = (Na_2SO_4 + Aq) + 2HCl,$$

whence we deduce the following proportion,

$$2\overset{117}{Na}Cl : 2\overset{73}{HCl} = 10 : x = \text{Ans. } 6.239 \text{ kilogrammes.}$$

Problem 2. It is required to calculate how much sulphuric acid and nitre must be used to make 250 grammes of the strongest nitric acid. The reaction is expressed by the equation

$$KNO_3 + H_2SO_4 = K, HSO_4 + HNO_3,$$

whence we get the proportions

$$\overset{63}{HNO_3} : \overset{98}{H_2SO_4} = 250 : x = \text{Ans. 1. } 388.9 \text{ grammes sulphuric acid.}$$

$$\overset{63}{HNO_3} : \overset{101.1}{KNO_3} = 250 : x = \text{Ans. 2. } 401.2 \text{ grammes nitre.}$$

The student should also solve by the same rule the problems given in the last section.

25. Gay-Lussac's Law. — This eminent French chemist was the first to state clearly the important truth, that, when gases or vapors react on each other, the volumes both of the factors and of the products of the reaction always bear to each other some very simple numerical ratio. This truth is generally known as the law of Gay-Lussac, but, since the principle is a direct consequence of the atomic theory, it is best studied in that relation. It is, as we have seen, a fundamental postulate of the theory that equal volumes of all substances, when in the aeriform condition, contain the same number of molecules. Hence it follows, that the volumes of all single molecules are the same, and, if we take this common volume as our unit of measure, it follows, further, that the total molecular volume represented by any symbol is always equal to the number of molecules. We are thus led to a most important fact, which gives an additional meaning to our chemical symbols, for it appears that *Every chemical equation, when properly written, represents not only the relative weights, but also the relative volumes of its factors and products, when in the state of gas.*

This principle is illustrated by the following equations:

$$\boxed{CH_4} + 2\boxed{O=O} = \boxed{CO_2} + 2\boxed{H_2O}$$
Marsh Gas.   Oxygen Gas.   Carbonic Anhydride.   Aqueous Vapor.

$$2\boxed{NO} + 5\boxed{H\text{-}H} = 2\boxed{NH_3} + 2\boxed{H_2O}$$
Nitric Oxide Gas.   Hydrogen Gas.   Ammonia Gas.   Aqueous Vapor.

The squares which here serve to indicate equal volumes, and to impress on the mind the meaning of the symbols, are evidently unnecessary and will not be used hereafter.

It is a great advantage of the crith, which has been proposed as a unit of weight in chemistry (see § 2), that it stands in the same relation to the French unit of volume, the litre, in which the weight of the hydrogen molecule stands to the common volume of all molecules, the unit of molecular volume. The volume of a given mass of any gas is always equal to the weight in criths divided by its specific gravity referred to hydrogen [3]. In like manner, the molecular volume represented by any symbol is equal to one half (§ 17) the total atomic weight of that symbol divided by the specific gravity of the gas it represents. In other words, the weight in criths of any mass of gas stands in the same relation to its volume in litres as that in which one half the total atomic weight of any symbol stands to the total molecular volume, or, what amounts to the same thing, to the number of molecules which the symbol represents. This relation must, of course, hold in every chemical equation, and hence, with a simple modification, the rule of the last section may be extended to the many cases in which it is desired to calculate the volumes of gas or vapor involved in a chemical change. The rule so modified may be stated thus : —

Express the reaction in an equation; make then the proportion, As one half of the symbol of the first substance is to the number of molecules of the second, so is the weight in criths of the first to the volume in litres of the second ; reduce the symbol to numbers, and calculate the value of the unknown quantity.

This rule has the same general application as the first, and a few examples will illustrate the use of it.

Problem 1. How much chlorate of potash must be used to obtain one litre of oxygen gas? The reaction is expressed by the equation

$$2KClO_3 = 2KCl + 3 O{=}O,$$

whence we get the proportion

$$\tfrac{1}{2}(2\overset{122.6}{KClO_3}) : 3 = x : 1. \quad x = 40.9 \text{ criths},$$

$$40.9 \times 0.0896 = \text{Ans. } 3.664 \text{ grammes}.$$

Problem 2. How many litres of oxygen gas can be obtained from 500 grammes of chlorate of potash? The reaction is the same as before, but in this case the grammes must first be reduced to criths. The proportion will then be written

$$\overset{122.6}{KClO_3} : 3 = \frac{500}{0.0896} : x = \text{Ans. } 136.6 \text{ litres}.$$

50 STOCHIOMETRY.

Problem 3. How many litres of ammonia gas NH_3 are contained in 20 grammes of ammonic chloride, $NH_3\text{-}HCl$? Here we require no equation; for the symbol itself gives at once the proportion

$$\tfrac{1}{2}NH_4Cl : 1 = \frac{20}{0.0896} : x = \text{Ans. 8.343 litres.}$$

In applying the rules of this chapter to the solving of stochiometrical problems, the student should carefully bear in mind, *first*, that the rule of (24) applies to all those cases in which the *weight* of one substance is to be calculated from the *weight* of another; *secondly*, that when *volume* is to be deduced from *volume* the answer can be found by mere inspection of the equation according to the principles stated in (25), and *thirdly*, that the rule of page 49 applies only to those problems in which volume is to be calculated from weight, or the reverse. In using this last rule it must be remembered that the "first substance" is always the one whose *weight* is given or sought, while the "second substance" is always the one whose *volume* is given or sought.

Questions and Problems.

1. What is the molecular weight of plumbic sulphate, $Pb\text{=}O_2\text{=}SO_4$? Of calcic phosphate, $Ca_3\overline{=}O_6\overline{=}(PO)_2$? Of ammonia alum, $(NH_4)_2, (Al_2)\overline{=}O_6\text{-}(SO_3)_4, 24H_2O$? Ans. 303, 310, and 906.8.

2. What are the molecular weights of the symbols

$$3C_2H_4O_2; \ 5(FeSO_4 \cdot 7H_2O) \text{ and } 7K_2\text{=}O_2\text{=}CO?$$

Ans. 180, 1390, and 967.4.

3. Are the total atomic weights of the two members of the following reaction equal?

$$Fe + (H_2SO_4 + Aq) = (FeSO_4 + Aq) + H\text{-}H.$$

Ans. The total weight of each member of the equation is 154.

4. Calculate the percentage composition of ammonic chloride, NH_4Cl. Ans. Nitrogen, 26.17; Hydrogen, 7.48; Chlorine, 66.35.

• 5. Calculate the percentage composition of nitrobenzoel, $C_6H_5NO_2$.
Ans. Carbon, 58.53; Hydrogen, 4.07; Nitrogen, 11.39; Oxygen, 26.01.

STOCHIOMETRY.

6. Given the percentage composition of chloroform as follows: Carbon, 10.04; Hydrogen, 0.83; Chlorine, 89.13. Required the symbol, knowing that the Sp. Gr. of chloroform vapor equals 59.75.

Ans. $CHCl_3$.

7. Given the percentage composition of stanno-diethylic bromide as follows: Tin, 35.13; Carbon, 14.29; Hydrogen, 2.97; Bromine, 47.61. Required the symbols, knowing that the Sp. Gr. of the vapor equals 168. Ans. $SnC_4H_{10}Br_2$.

8. Given the percentage composition of ethylene chloride as follows: Carbon, 24.24; Hydrogen, 4.04; Chlorine, 71.72. Required the symbol, knowing that the Sp. Gr. of the vapor equals 49.5.

Ans. $C_2H_4Cl_2$.

9. Given the percentage composition of cream of tartar as follows: Potassium, 20.79; Hydrogen, 2.66; Carbon, 25.52; Oxygen, 51.03. Required the simplest symbol possible. Ans. $KH_5C_4O_6$.

10. Given the percentage composition of crystallized ferrous sulphate as follows: Iron, 20.15; Sulphur, 11.51; Oxygen, 23.02; Water, 45.32. Required the simplest symbol possible.

Ans. Estimating the number of molecules of water (H_2O), as if water were a fourth element with an atomic weight of 18, we get $FeSO_4 \cdot 7H_2O$.

11. The percentage composition of morphia according to Liebig's analysis is Carbon, 71.35; Hydrogen, 6.69; Nitrogen, 4.99; Oxygen (by loss), 16.97. What is the symbol of this alkaloid, and how closely does this symbol agree with the results of analysis?

Ans. The symbol $C_{17}H_{19}NO_3$ would require 71.58 Carbon, 6.66 Hydrogen, 4.91 Nitrogen, and 16.85 Oxygen.

12. It is required to find the weight of phosphorus in 155 kilos. of calcic phosphate ($Ca_3P_2O_8$). Ans. 31 kilos.

13. It is required to find the weight of sulphuric anhydride (SO_3) in 284 kilos. of sodic sulphate, Na_2SO_4. Ans. 80 kilos.

14. How many grammes of plumbic sulphate ($PbSO_4$) can be made from 2.667 grammes of sulphuric anhydride (SO_3)?

Ans. 30.3 grammes.

15. How many grammes crystallized cupric sulphate ($CuSO_4 \cdot 6H_2O$) will yield 317 grammes of copper? Ans. 1337 grammes.

16. Required the total molecular weight of crystallized sodic phosphate, knowing that 71.6 parts of the salt contain 9.2 parts of sodium, and that the total atomic weight of sodium in one molecule of the compound is 46. Ans. 358.

17. The molecular weight of potassic nitrate is 101.1, and 2.359 grammes of the salt contain 1.120 grammes of oxygen. What is the total atomic weight of oxygen, and also the number of oxygen atoms in one molecule?

Ans. Total atomic weight 48. No. of oxygen atoms 3.

18. How much nitric acid (HNO_3) is required to dissolve 3.804 grammes of copper (Cu) and how much cupric nitrate (CuN_2O_6) and how much nitric oxide (NO) will be formed in the process? The reaction is expressed by the equation

$$3Cu + (8HNO_3 + Aq) = (3CuN_2O_6 + 4H_2O + Aq) + \mathbb{NO}.$$

Ans. 10.08 grammes of nitric acid; 11.244 grammes of cupric nitrate and 0.60 grammes of nitric oxide.

19. How much common salt ($NaCl$) must be added to a solution containing 30 grammes of argentic nitrate ($AgNO_3$) in order to throw down the whole of the silver, and how much argentic chloride ($AgCl$) will be thus precipitated?

$$(AgNO_3 + NaCl + Aq) = \mathbf{AgCl} + (NaNO_3 + Aq).$$

Ans. 9.75 grammes of salt and 23.92 grammes argentic chloride.

20. How many litres of ammonia gas ($\mathbb{NH}_3$) and how many of chlorine gas $\mathbb{Cl}$-$\mathbb{Cl}$ are required to make one litre of nitrogen gas $\mathbb{N}$=$\mathbb{N}$? How many litres of hydrochloric acid gas ($\mathbb{HCl}$) are also formed?

$$2\mathbb{NH}_3 + 3\mathbb{Cl}\text{-}\mathbb{Cl} = 6\mathbb{HCl} + \mathbb{N}\text{=}\mathbb{N}.$$

Ans. 2 litres of ammonia gas; 3 litres of chlorine gas, and 6 litres of hydrochloric acid gas.

21. How many litres of hydrochloric acid gas ($\mathbb{HCl}$) and how many of oxygen gas ($\mathbb{O}$=$\mathbb{O}$) can be obtained from one litre of aqueous vapor ($\mathbb{H}_2\mathbb{O}$), and how many litres of chlorine gas ($\mathbb{Cl}$-$\mathbb{Cl}$) must be used in the process?

$$2\mathbb{H}_2\mathbb{O} + 2\mathbb{Cl}\text{-}\mathbb{Cl} = 4\mathbb{HCl} + \mathbb{O}\text{=}\mathbb{O}.$$

Ans. 2 litres of hydrochloric acid gas, ½ litre of oxygen gas, and 1 litre of chlorine gas.

22. How many litres of oxygen gas ($\mathbb{O}$=$\mathbb{O}$) are required to burn completely (i. e. to combine with) one litre of alcohol vapor ($\mathbb{C}_2\mathbb{H}_6\mathbb{O}$), and how many litres of carbonic anhydride ($\mathbb{CO}_2$) and how many of aqueous vapor ($\mathbb{H}_2\mathbb{O}$) are formed by the process? The chemical reaction which takes place when alcohol burns is expressed by the equation

STOCHIOMETRY.

$$C_2H_6O + 3 O{=}O = 2 CO_2 + 3 H_2O.$$

Ans. 3 litres of oxygen gas; 2 litres of carbonic anhydride, and 3 litres of aqueous vapor.

23. How many litres of oxygen gas are required to burn one litre of arseniuretted hydrogen (H_3As), and how many litres of arsenious acid vapor (AsO_3) and how many of aqueous vapor are formed in the process?

$$4 H_3As + 9 O{=}O = 4 AsO_3 + 6 H_2O.$$

Ans. 2¼ litres of oxygen gas; 1 litre arsenious acid vapor and 1½ litres of aqueous vapor.

24. How many litres of chlorine gas can be made with 19.49 grammes of manganic oxide (MnO_2)?

$$\mathbf{MnO_2} + (4HCl + Aq) = (MnCl_2 + 2H_2O + Aq) + Cl{-}Cl.$$

Ans. 5 litres.

25. How many grammes of chalk ($CaCO_3$) are required to yield one litre of carbonic anhydride?

$$\mathbf{CaCO_3} + (2HCl + Aq) = (CaCl_2 + H_2O + Aq) + CO_2.$$

Ans. 4.48 grammes.

26. How many litres of hydrochloric acid gas (HCl) can be made with 8.177 kilogrammes of common salt ($NaCl$)?

$$(2NaCl + H_2SO_4 + Aq) = (Na_2SO_4 + Aq) + 2 HCl.$$

Ans. 5000.

27. How many grammes of ferrous sulphide (FeS) are required to yield 568 $\overline{c.m.}^3$ of sulphuretted hydrogen (H_2S)?

$$\mathbf{FeS} + (H_2SO_4 + Aq) = (FeSO_4 + Aq) + H_2S.$$

Ans. 2.24 grammes.

CHAPTER VII.

CHEMICAL EQUIVALENCY.

26. *Chemical Equivalents.* — If in a solution of argentic sulphate we place a strip of metallic copper, we find after a short time that all the silver has separated from the solution, and that a certain quantity of copper has dissolved in its place.

$$(Ag_2SO_4 + Aq) + \mathbf{Cu} = (CuSO_4 + Aq) + \mathbf{Ag_2}. \quad [25]$$

If now we pour off the solution of cupric sulphate, and place in this solution a strip of metallic zinc, the metallic copper in its turn will all separate, and to replace it a certain amount of zinc will dissolve.

$$(CuSO_4 + Aq) + \mathbf{Zn} = (ZnSO_4 + Aq) + \mathbf{Cu}. \quad [26]$$

Lastly, if we pour off the solution of zincic sulphate, and place in this a strip of metallic magnesium, the zinc will in like manner be replaced by magnesium.

$$(ZnSO_4 + Aq) + \mathbf{Mg} = (MgSO_4 + Aq) + \mathbf{Zn}. \quad [27]$$

In experiments like these, we can by proper analytical methods determine the relative quantities by weight of the several metals which thus replace each other, and we find that they are always the same. Thus, if our first solution contained 108 milligrammes of silver, the amount of each metal successively dissolved and precipitated would be, of copper, 31.7 $m.\,g.$, of zinc, 32.6 $m.\,g.$, of magnesium, 12 $m.\,g.$ Moreover, if, instead of using in our experiments a metallic sulphate, we take a metallic chloride, nitrate, acetate, or any other compound of the metals, we find that the same definite ratios are preserved, at least in every case where the substitution is possible. It would appear then that these relative quantities of the several metals exactly replace each other in all such cases. They are, therefore, regarded as the chemical equivalents of each other, in the sense that they are capable of filling each other's place.

In a strict sense, two quantities of different elements can be said to be equivalent to each other only when they are actually capable of replacing each other in some known chemical reaction, but formerly the word was used with a much wider significance, and quantities of two different elements were said to be equivalent to each other if they had been proved to be equivalent to the same quantity of some third element which served as a link of connection. In this way an equivalency may be established between all the chemical elements, and the system of chemistry still used in many textbooks is based on a system of equivalency so determined. If the table of chemical equivalents on this old system is compared with a table of atomic weights on the new, it will be found that the numbers of the one are either the same as those of the other, or else some very simple multiples of them. The one set of numbers can be used in all stochiometrical calculations in the same way as the other, and on the old system the symbols stand for equivalents, as in the new they stand for atomic weights. The equivalents have this advantage, that they are the result of direct experiments, and are based on no hypothesis in regard to the molecular constitution of matter. But this hypothesis is necessary, in order to correlate a large number of facts which modern chemical investigation has brought to light, and when once made, the rest of the system follows as a necessary consequence.

27. *Quantivalence and Atomicity of the Elements.* — If now, starting with the atomic weights as they have been determined or assumed in Table II., we compare together the different elements from the point of view taken in the last section, it will be found, that, while in some cases one atom of one element is the equivalent of one atom of another in other cases, it may be the equivalent of two, three, or four atoms. Since in the system of this book the symbols always stand for atomic weights, the relation here referred to is made evident whenever any metathetical reaction is expressed in the form of an equation. A few examples will illustrate the point, and make clear what is meant. The reaction of aqueous hydrochloric acid on a solution of argentic nitrate is expressed by the

$$(\overset{\text{I}}{Ag}NO_3 + HCl + Aq) = (HNO_3 + Aq) + \textbf{AgCl}, \quad [28]$$

and here evidently Ag changes places with H, and hence one atom of silver is equivalent to one atom of hydrogen. Take now the reaction of dilute sulphuric acid on zinc, which is expressed by the equation,

$$\overset{\text{II}}{Zn} + (H_2SO_4 + Aq) = (ZnSO_4 + Aq) + \text{H-H}, \quad [29]$$

and it will be seen that Zn has changed places with H_2, and hence that one atom of zinc is the equivalent of two atoms of hydrogen. Lastly, in the reaction of water on phosphorous trichloride, expressed by the equation,

$$H_3H_3O_3 + \overset{\text{III}}{P}Cl_3 = 3HCl + \underset{\text{Phosphorous Acid.}}{H_3PO_3}, \quad [30]$$

it is equally evident that P has changed places with H_3, and hence in this reaction one atom of phosphorus is equivalent to three atoms of hydrogen.

This relation of the elements to each other is called by Hofmann *quantivalence*; and selecting here, as in the system of atomic weights, the hydrogen atom as our standard of reference, the atoms of different elements are called *uni*valent, *bi*valent, *tri*valent, or *quadri*valent, according as they are in the sense already indicated *equi*valent to one, two, three, or four atoms of hydrogen. These terms are very appropriate, since they are all derived from the same root as our common English word equivalent, which best expresses the fundamental idea that underlies the whole subject. We shall therefore adopt them in this book, and, as Hofmann recommends, designate the quantivalence, whenever important, by a Roman numeral placed over the atomic symbol thus,

$$\overset{\text{I}}{Cl}, \quad \overset{\text{II}}{O}, \quad \overset{\text{III}}{N}, \quad \overset{\text{IV}}{C}.$$

In most cases, however, the quantivalence is indicated with sufficient clearness by the dashes, which are also used in this book to separate the parts of a molecular symbol. The number of these dashes is always the same as the quantivalence of the atoms, or groups of atoms, on either side.

With these additions to our notation we are able to express by our symbols all that was valuable in the old system of equivalents, and at the same time all that is peculiar to our modern theories.

Precisely the same relations of quantivalence are manifested even more fully by the compound radicals, whenever in a chemical reaction they change places with elementary atoms, and their replacing value is indicated in the same way. Thus, in the following reaction,

$$\underset{\text{Acetyl chloride.}}{C_2H_3\overset{\text{I}}{O}\text{-}Cl} + \underset{\text{Water.}}{H\text{-}O\text{-}H} = H\text{-}Cl + \underset{\text{Acetic Acid.}}{H\text{-}O\text{-}C_2H_3\overset{\text{I}}{O}}, \quad [31]$$

the radical C_2H_3O, named acetyl, changes places with one atom of hydrogen, and is therefore univalent, while in the next,

$$\underset{\text{Chloroform.}}{\overset{\text{III}}{C}H \equiv Cl_3} + H_3N = 3HCl + \underset{\text{Hydrocyanic Acid.}}{\overset{\text{III}}{C}H \equiv N}, \quad [32]$$

the radical CH is as evidently trivalent.

The quantivalence of an element or radical is shown, not only by its power of replacing hydrogen atoms, but also by its power of replacing any other atoms whose quantivalence is known. Moreover, what is still more important, the quantivalence of an element or radical is shown, not only by its replacing power, but also by what we may term its *atom-fixing power*, that is, by its power of holding together other elements or radicals in a molecule. We may take as examples the molecules of four very characteristic compounds, namely, hydrochloric acid, water, ammonia, and marsh gas, whose symbols may be written thus,

$$\underset{\text{Hydrochloric Acid.}}{\overset{\text{I}}{H}\text{-}Cl} \qquad \underset{\text{Water.}}{H, \overset{\text{II}}{H} \text{=} O} \qquad \underset{\text{Ammonia.}}{H, H, \overset{\text{III}}{H} \text{=} N} \qquad \underset{\text{Marsh Gas.}}{H, H, H, \overset{\text{IV}}{H} \equiv C}.$$

By these symbols it appears, that, while the univalent atom of chlorine can hold but one atom of hydrogen, the bivalent atom of oxygen holds two, the trivalent atom of nitrogen three, and the quadrivalent atom of carbon four atoms of the same element. It appears, then, that the Roman numerals or dashes, which represent the replacing power of the atoms or radicals, represent also the *atom-fixing power* of the same, measured in each case by the number of atoms of hydrogen, or their equivalents, with which these atoms or radicals can combine to form a single molecule. On account of the importance of this principle we will extend our illustrations to a number of other compounds, and the student should carefully compare in each case the quantivalence on the two sides of the dash or dashes, which mark the atom-fixing power of the dominant atom in the molecule.

$$\overset{\text{I}}{Na}\text{-}\overset{\text{I}}{Cl} \qquad \overset{\text{I}}{K}\text{-}\overset{\text{I}}{I} \qquad \overset{\text{I}}{C_2H_5}\text{-}\overset{\text{I}}{Br} \qquad \overset{\text{I}}{K}\text{-}\overset{\text{I}}{CN};$$
Sodic Chloride. Potassic Iodide. Ethylic Bromide. Potassic Cyanide.

$$\overset{\text{I}}{K}\text{-}\overset{\text{II}}{O}\text{-}\overset{\text{I}}{H} \qquad \overset{\text{II}}{Pb}\text{=}\overset{\text{II}}{O} \qquad \overset{\text{I}}{H}\text{-}\overset{\text{II}}{O}\text{-}\overset{\text{I}}{NO_2} \qquad \overset{\text{I}}{H}\text{-}\overset{\text{II}}{O}\text{-}\overset{\text{I}}{C_2H_3O};$$
Potassic Hydrate. Plumbic Oxide. Nitric Acid. Acetic Acid.

$$\overset{\text{I}}{H}, \overset{\text{I}}{H}, \overset{\text{I}}{C_2H_5}\overset{\text{III}}{\equiv}N \qquad (\overset{\text{I}}{C_2H_5})_3\overset{\text{III}}{\equiv}P \qquad \overset{\text{I}}{CH_3}, \overset{\text{I}}{C_2H_5}, \overset{\text{I}}{C_5H_{11}}\overset{\text{III}}{\equiv}N.$$
Ethylamine. Triethyl phosphine. Methyl-ethyl-amyl-amine.

The quantivalence of the chemical elements, especially as indicated by their atom-fixing power, is by no means always the same. They constantly exhibit under different conditions an unequal atom-fixing power. Thus we have

$$\overset{\text{II}}{Sn}Cl_2 \text{ and } \overset{\text{IV}}{Sn}Cl_4, \quad \overset{\text{III}}{P}Cl_3 \text{ and } \overset{\text{V}}{P}Cl_5, \quad \overset{\text{III}}{N}H_3 \text{ and } \overset{\text{V}}{N}H_4Cl.$$

Each element, however, has a maximum power, which it never exceeds. This we shall call its *atomicity*, and we shall distinguish the elements as monads, dyads, triads, &c., according to the number of univalent atoms or radicals they are able at most to bind together. Thus nitrogen is a pentad, although it is more commonly trivalent, and lead is a tetrad, although it is usually bivalent. Again, sulphur is a hexad, although in most of its relations it is, like lead, bivalent. In like manner with other elements, one of the few possible conditions is generally much more common and stable than the rest, and this prevailing quantivalence of an element is a more characteristic property than its maximum quantivalence or atomicity. A classification of the elements based on their atomicity alone would contravene their most striking analogies, while one based on the prevailing quantivalence very nearly satisfies all natural affinities. Moreover, it should be added, that, while the prevailing quantivalence of the elements is generally well established, their atomicity is frequently still in doubt; for the first can generally be discovered by studying the simple compounds of the elements with chlorine or hydrogen, while the last is often only manifested in those more complex combinations, in regard to which a difference of opinion is possible.

The possible degrees of quantivalence of an elementary atom are related to each other by a very simple law. They are either all even or all odd. Thus the atom of sulphur may be sextivalent, quadrivalent and bivalent, but is never trivalent or univalent; and on the other hand the atom of nitrogen may be quinquivalent, trivalent and univalent, but not quadrivalent or bivalent. Atoms like those of sulphur, whose quantivalence is always even, are called *artiads*, while those like nitrogen, whose quantivalence is always odd, are called *perissads*.

28. *Atomicity or Quantivalence of Radicals.* — When in the molecule of any compound the dominant or central atom is united to as many other atoms as it can hold of that kind, the molecule is said to be saturated; thus

$$HCl, \quad H_2O, \quad H_3N, \quad H_4C$$

are all saturated molecules; for, although nitrogen is a pentad, it cannot without the intervention of some other atom or radical hold more than three atoms of hydrogen. While on the other hand the molecules

$$\overset{\text{II}}{C}O, \quad \overset{\text{II}}{P}Cl_3 \text{ and } \overset{\text{II}}{Sn}Cl_2$$

are not saturated, for they can combine directly with more oxygen or chlorine, forming thus the saturated molecules

$$CO_2, \quad PCl_5 \text{ and } SnCl_4.$$

If now from a saturated molecule we withdraw one or more atoms of hydrogen, or their equivalents, *the residue may be regarded as a compound radical with an atomicity equal to the number of hydrogen atoms, or their equivalents, withdrawn.* Thus, if from the saturated molecule of marsh gas H_4C we withdraw one atom of hydrogen, we get the radical methyl H_3C, which is a monad; if we withdraw two atoms, we have the radical, H_2C, which is a dyad; if we withdraw three, there results HC, which is a triad; and lastly, if we withdraw all four, we fall back on the tetrad atom of carbon. Again, if from the saturated molecule of nitric anhydride N_2O_5 we withdraw one atom of the dyad oxygen O, it falls into two atoms of NO_2 each of which is a monad. If now we withdraw from NO_2 one of its remaining atoms of oxygen, we have left NO, which is a triad. Lastly, a molecule of sulphuric anhydride SO_3, which is saturated, gives, by withdrawing one atom of oxygen, SO_2, which acts as a bivalent radical. These considerations lead us to a simple rule, first stated by Wurtz, which in almost every case will enable us to infer the atomicity of any given radical. *The atomicity[1] of a compound radical is always equal to the number of hydrogen atoms, or their equivalents, which the radical may be regarded as having lost.*

It must not be supposed, however, that all such radicals are possible compounds. In a few cases only these residues, of which we have been speaking, form non-saturated molecules, which are capable of existing in a free state, like those of carbonic oxide, nitric oxide and sulphurous acid. At other times they are compound radicals, which, *by doubling*, form molecules that can exist in a free state, as those of cyanogen gas, and perhaps also of some hydrocarbons. Again, they appear as compound radicals, which pass and repass in so many chemical reactions as to almost force upon us the belief that they have a real existence, and represent the actual grouping of the atoms in the compounds of which they seem to be an integral part. Still again, and even more frequently, they can only be regarded as convenient factors in a chemical equation.

[1] The quantivalence of a compound radical is always the same as its atomicity.

CHEMICAL EQUIVALENCY. 61

Questions and Problems.

1. Analyze the following metathetical reactions, showing in each case how many parts of the several elements are equivalent to one part by weight of hydrogen, and also to how many atoms of hydrogen one atom of each of the interchanging elements corresponds. For the atomic weights refer to Table II.

$$2H\text{-}O\text{-}C_2H_5 + K\text{-}K = 2K\text{-}O\text{-}C_2H_5 + H\text{-}H.$$
Alcohol. Potassium. Potassic Ethylate.

$$2H\text{-}O\text{-}H + Mg = Mg\text{=}O_2\text{=}H_2 + H\text{-}H.$$
Water. Magnesic Hydrate.

$$Sb\equiv O_3\equiv H_3 + 3HCl = SbCl_3 + 3H\text{-}O\text{-}H.$$
Antimonious Hydrate. Antimonious Chloride.

$$4H\text{-}O\text{-}H + SiCl_4 = H_4\equiv O_4\equiv Si + 4HCl.$$
Silicic Chloride. Silicic Acid.

2. Make out a table of chemical equivalents so far as the reactions of this chapter will enable you to deduce them from the atomic weights given in Table II.

3. Analyze the following metathetical reactions, showing in each case how the quantivalence of the several compound radicals involved in the metathesis, is indicated.

$$H\text{-}O\text{-}H + (C_2H_3O)\text{-}O\text{-}(C_2H_5) = (C_2H_3O)\text{-}O\text{-}H + H\text{-}O\text{-}(C_2H_5).$$
Water. Acetic Ether. Acetic Acid. Alcohol.

$$2K\text{-}(CN) + (C_2H_4)\text{=}Br_2 = (C_2H_4)\text{=}(CN)_2 + 2KBr.$$
Potassic Cyanide. Ethylene Bromide. Ethylene Cyanide. Potassic Bromide.

$$3H\text{-}O\text{-}H + (C_3H_5)\equiv Cl_3 = (C_3H_5)\equiv O_3\equiv H_3 + 3HCl.$$
Water. Glyceryl Chloride. Glycerine. Hydrochloric Acid.

The names of the radicals are as follows: C_2H_3O, Acetyl; C_2H_5, Ethyl; C_2H_4, Ethylene; C_3H_5, Glyceryl; CN, Cyanogen.

4. What is the atom-fixing power or quantivalence of the different atoms and radicals in the following symbols?

$$K_3\equiv S_3\equiv SbS$$
Potassic Sulphantimonite.

$$H,Na = O_2\text{=}CO$$
Acid Sodic Carbonate.

$$(NH_4)\text{-}O\text{-}NO$$
Ammonic Nitrite.

$$H_4\equiv N_2\text{=}C_2O_2$$
Oxamide.

$$(HO),(H_2N)\text{=}(C_4H_4O_2)$$
Succinamic Acid.

$$K,Sb\equiv O_4\equiv C_4H_2O_2.$$
Tartar Emetic (dried).

5. If H_2O; C_2H_6; $C_2H_6O_2$ (alcohol); $COCl_2$ (phosgene gas); $C_2H_4O_2$ (acetic acid) and $C_2H_2O_4$ (oxalic acid) are saturated molecules, what is the atomicity of the radicals HO (hydroxyl); C_2H_5 (ethyl); C_2H_4 (ethylene); C_2H_6O (aldehyde); CO (carbonyl); C_2H_3O (acetyl) and C_2O_2 (oxalyl).

CHAPTER VIII.

CHEMICAL TYPES.

29. *Types of Chemical Compounds.*—There are three modes or forms of atomic grouping, to which so large a number of substances may be referred, that they are regarded as molecular types, or patterns, according to which the atoms of a molecule are grouped together. These types may be represented by the general formulæ:—

$$\overset{\text{I}}{R}\text{-}\overset{\text{I}}{R} \qquad \overset{\text{I}}{R},\overset{\text{II}}{R}\text{=}\overset{\text{I}}{R} \quad \text{or} \quad \overset{\text{I}}{R}\text{-}\overset{\text{II}}{R}\text{-}\overset{\text{I}}{R} \qquad [33]$$
$$\overset{\text{I}}{R},\overset{\text{I}}{R},\overset{\text{I}}{R}\text{=}\overset{\text{III}}{R} \quad \text{or} \quad \overset{\text{I}}{R},\overset{\text{I}}{R}\text{=}\overset{\text{III}}{R}\text{-}\overset{\text{I}}{R}.$$

It will be noticed, that in the first of these types a single univalent atom or radical[1] is united to another single univalent atom, that in the second a bivalent atom binds together two univalent atoms or their equivalents, and that in the third a trivalent atom binds together three univalent atoms, or their equivalents. The dashes are used to separate what has been called the *central*, the *dominant*, or the *typical* atom from those which it thus unites into one molecular whole, and serve at the same time to point out the parts of the symbol to which its affinities are directed. Commas are used to separate the subordinate atoms so united. It will be further noticed, that in each case the quantivalence of the dominant atom is equal to the sum of the quantivalences of the subordinate atoms, or radicals, on either side; and the peculiarity in each case consists solely in the relations of the parts of the molecule which we thus attempt to indicate by the symbol. The three compounds, hydrochloric acid, water, and ammonia,

$$\overset{\text{I}}{H}\text{-}\overset{\text{I}}{Cl}, \quad \overset{\text{I}}{H},\overset{\text{I}}{H}\text{=}\overset{\text{II}}{O}, \quad \overset{\text{I}}{H},\overset{\text{I}}{H},\overset{\text{I}}{H}\text{=}\overset{\text{III}}{N},$$

---

[1] Here, as elsewhere through the book, we use the symbol $\overset{\text{I}}{R}$ for any univalent, $\overset{\text{II}}{R}$ for any bivalent, and $\overset{\text{III}}{R}$ for any trivalent atom or radical. Moreover, to avoid unnecessary repetition, we shall for the future conform to the general usage, and speak of the atoms of a radical as well as of those of an element, and use the word "atom" as applying to both, although the usage frequently involves an obvious solecism.

are generally taken as representatives of these types, and substances are described as belonging to the type of hydrochloric acid, to the type of water, or to the type of ammonia, as the case may be. These substances, however, are regarded as types in no other sense than that their molecules present the same mode of grouping which is indicated above by the more general symbols. Substances belonging to the same type may have widely different properties. To the type of water belong the strongest alkalies and the most corrosive acids known. In what, then, it may be asked, does the type outwardly consist, or in what is it manifested? for the grouping of the atoms can only be a matter of inference. The answer is, that the type of the molecules of a substance is manifested solely by its chemical reactions. Substances belonging to the same type are simply those whose reactions may be classed together according to some one general plan. Thus water, alcohol, and acetic acid are classed in the same type, because, when submitted to the action of the same or similar reagents, they undergo a like transformation, which seems to point to a similarity of atomic grouping.

$$H, H\text{-}O + PCl_5 = PCl_3O + H\text{-}Cl + H\text{-}Cl \cdot$$
Water.　　Phosphoric Chloride.　　　　　Hydrochloric Acid.

$$H, C_2H_5\text{-}O + PCl_5 = PCl_3O + H\text{-}Cl + C_2H_5\text{-}Cl \quad [34]$$
Alcohol.　　Phosphoric Oxz-chloride.　　　　Ethyl Chloride.

$$H, C_2H_3O\text{-}O + PCl_5 = PCl_3O + H\text{-}Cl + C_2H_3O\text{-}Cl.$$
Acetic Acid.　　　　　　　　　　　　　　　Acetyl Chloride.

On studying these reactions, it will be seen that both the manner in which the three compounds break up, and the probable constitution of the products formed, point to the conclusion, that, in each, one bivalent atom holds together two univalent atoms or radicals. It will be found, in the first place, that in all three cases the reaction consists primarily in the substitution of two atoms of chlorine for one of oxygen in the original molecule. It will appear, in the next place, that as soon as this dominant atom, which holds together the parts of the molecule, is taken away, each of the three molecules splits up into two others of a similar type; and lastly, it is evident from the third example that one of the oxygen atoms of acetic acid stands in a very different relation to the molecule from the other. All this points to the inference just made. At least, these and a vast number of similar reactions are best explained on this hypothesis, and herein its only value lies and its probability rests. In section 27 we have already given the symbols of a number of chemical compounds so printed that they can be at once referred to one or the other of the three types here alluded to, and it will not, therefore, be necessary to multiply examples in this place.

30. *Condensed Types.* — In the same way that a bivalent atom may bind together two univalent atoms or their equivalents, so, also, it may serve to bind together two *molecules*, and, in like manner, a trivalent atom may bind together three *molecules* into a more complex molecular group; and thus are formed what are called condensed types. We may represent a double molecule of the type of water thus, $\overset{I}{R_2}=\overset{II}{R_2}=\overset{I}{R_2}$, but it must be borne in mind that such a symbol stands for two molecules, since, by the very definition, two molecules of the same kind cannot chemically combine. We can, however, solder them, as it were, into one molecular whole by substituting for the two univalent atoms $\overset{I}{R_2}$ a single bivalent atom $\overset{II}{R}$, when we obtain a mode of molecular grouping represented by

$$\overset{I}{R_2}=\overset{II}{R_2}=\overset{II}{R}, \qquad [35]$$

which may be called the type of water doubly condensed. The constitution of common sulphuric acid is best represented after this type by the symbol, —

$$H_2=O_2=\overset{II}{S}O_2. \qquad [36]$$

The soldering atom is here the bivalent radical $\overset{II}{S}O_2$. In like manner, by using a trivalent atom, we can solder together three molecules of the same water-type, as in the general symbol, —

$$\overset{I}{R_3}=\overset{II}{R_3}=\overset{III}{R}, \qquad [37]$$

which represents the type of water trebly condensed. In the same way we may derive the symbol, —

$$\overset{I}{R_2}, \overset{I}{R_2}\equiv\overset{III}{R_2}=\overset{II}{R}, \qquad [38]$$

which represents the type of ammonia doubly condensed. The substance urea, one of the most important of the animal secretions, is best represented by a symbol after this last type,—

$$H_2, \; H_2 \overset{\text{III}}{\equiv} N_2 \text{-} \overset{\text{II}}{CO} \qquad [39]$$

where the soldering atom is the bivalent radical carbonyl.

Chemists have also been led to admit the existence of what are called *mixed types*, which are formed by the union of molecules of different types soldered together by a single multivalent atom or radical as before. Thus, the molecules of sulphurous acid may be regarded as formed of a molecule of water soldered to a molecule of hydrogen by an atom of sulphuryl, $\overset{\text{II}}{SO_2}$; thus, $H\text{-}O\text{-}H$ and $H\text{-}H$, united by $\overset{\text{II}}{SO_2}$ give

$$H\text{-}\overset{\text{II}}{O}\text{-}\overset{\text{II}}{SO_2}\text{-}H. \qquad [40]$$

So, also, the composition of a complex organic compound called sulphamide, or sulphamic acid, is most simply expressed when regarded as formed by the union of water and ammonia soldered together by the same radical sulphuryl; thus, from

$$H, \; H\overset{\text{III}}{=}N\text{-}H, \text{ and } H\text{-}\overset{\text{II}}{O}\text{-}H \text{ we have } H, \; H\overset{\text{III}}{=}N\text{-}\overset{\text{II}}{SO_2}\text{-}\overset{\text{II}}{O}\text{-}H. \qquad [41]$$

Lastly, if we bind together on the same principle molecules of the type of hydrochloric acid, we shall simply reproduce the types of water and of ammonia, thus showing that all the types are only condensed forms of the simplest. We must not, therefore, attach to the idea of a chemical type any deeper significance than that indicated above. It is simply a convenient mode of classifying certain groups of chemical reactions, and a help in representing them to the mind; and we may regard the same substance as formed on one type or on the other, as will best help us to explain the reactions we are studying. Moreover, it is frequently convenient to assume other types besides those here specially mentioned.

31. *Substitution.* — When cotton-wool is dipped in strong nitric acid (rendered still more active by being mixed with twice its volume of concentrated sulphuric acid), and afterwards washed and dried, it is rendered highly explosive, and,

although no important change has taken place in its outward aspect, it is found on analysis to have lost a certain amount of hydrogen and to have gained from the nitric acid an equivalent amount of nitric peroxide NO_2 in its place.

$$C_6(H_{10})O_5 \text{ becomes } C_6\big(H_7(NO_2)_3\big)O_5.$$
Cotton.   Gun-Cotton.

Under the same conditions glycerine undergoes a like change, and is converted into the explosive nitro-glycerine, —

$$C_3(H_8)O_3 \text{ becomes } C_3\big(H_5(NO_2)_3\big)O_3.$$
Glycerine.   Nitro-glycerine.

So, also, the hydrocarbon naphtha, called benzole, is changed into nitro-benzole, —

$$C_6H_6 \text{ becomes } C_6(H_5,NO_2).$$
Benzole.   Nitro-benzole.

The last compound is not explosive, and the explosive nature of the first two is in a measure an accidental quality, and is evidently owing to the fact that into an already complex structure there have been introduced, in place of the indivisible atoms of hydrogen, the atoms of a highly unstable radical rich in oxygen. The point of chief interest for our chemical theory is that this substitution does not alter, at least profoundly, the outward aspect of the original compound. Every one knows how closely gun-cotton resembles cotton-wool. In like manner nitro-glycerine is an oily liquid like glycerine, and nitro-benzole, although darker in color, is a highly aromatic volatile fluid like benzole itself. Products like these are called *substitution products*, and they certainly suggest the idea that each chemical compound has a certain definite structure, which may be preserved even when the materials of which it is built are in part at least changed. If in the place of firm iron girders we insert weak wooden beams, a building, while retaining all its outward aspects, may be rendered wholly insecure, and so the explosive nature of the products we have been considering is not at all incompatible with a close resemblance, in outward aspects and internal structure, to the compounds from which they were derived.

The idea that each body has a definite atomic structure is

even more forcibly suggested by another class of substitution products first studied by Dumas, in which atoms of chlorine, bromine, or iodine have taken the place of the hydrogen atoms of the original compound. Thus, if we act upon acetic acid with chlorine gas, we may obtain three successive products, as shown in the following table, although only the first and the last have been fully investigated.

Acetic acid	$C_2H_4O_2$	or	$(C_2H_3\overset{I}{O})\text{-}\overset{II}{O}\text{-}\overset{I}{H}$
Chloracetic acid	$C_2(H_3Cl)O_2$	"	$(C_2H_2Cl\overset{I}{O})\text{-}\overset{II}{O}\text{-}\overset{I}{H}$
Dichloracetic acid	$C_2(H_2Cl_2)O_2$	"	$(C_2HCl_2\overset{I}{O})\text{-}\overset{II}{O}\text{-}\overset{I}{H}$
Trichloracetic acid	$C_2(HCl_3)O_2$	"	$(C_2Cl_3\overset{I}{O})\text{-}\overset{II}{O}\text{-}\overset{I}{H}$

We cannot, however, replace the fourth atom of hydrogen by chlorine; and this fact seems to prove that there is a real difference between this atom of hydrogen and the other three, and gives an additional ground for the distinction we make when we write the symbol of acetic acid after the type of water, as in the second column. The three atoms of hydrogen in the radical placed on the left-hand side of the dominant atom may all be replaced by chlorine, but the single atom of hydrogen placed on the right cannot. These products all resemble acetic acid in that they form with the alkalies crystalline salts, when the fourth atom of hydrogen is replaced by an atom of sodium or potassium, as the case may be.

It was the study of these and similar substitution products which first led to the conception of *chemical types*, and the word as first used was intended to convey the idea of a definite structure, although perhaps as yet unknown; but as the theory was extended more and more, and to widely different chemical compounds, it was found that the first definite conception could not be maintained, and the idea gradually assumed the shape we have given it in the last section. Still, the facts from which the original conception was drawn remain, and they point no less clearly now than they did before to the existence of a definite structure in all chemical compounds as the legitimate object of chemical investigation.

32. Isomorphism. — Closely associated with the facts of the last section, which find their chief manifestation in substances of organic origin, are the phenomena of isomorphism, which are equally conspicuous among artificial salts and native minerals. There seems to be an intimate connection between chemical composition and crystalline form, and two substances which under a like form have an analogous composition are said to be *isomorphous*. Thus the following minerals all crystallize in rhombohedrons (Fig. 1,) which have very nearly the same interfacial angles, and, as the symbols show, they have an analogous composition. They are therefore isomorphous.

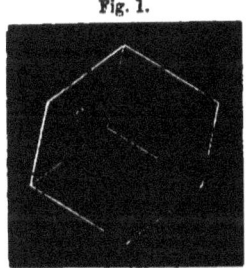

Fig. 1.

Calcite or calcic carbonate $\quad Ca\overset{\shortmid\shortmid}{=}O_2\overset{\shortmid\shortmid}{=}\overset{\shortmid\shortmid}{C}O$

Magnesite or magnesic carbonate $\quad Mg\overset{\shortmid\shortmid}{=}O_2\overset{\shortmid\shortmid}{=}\overset{\shortmid\shortmid}{C}O$

Chalybdite or ferrous " $\quad Fe\overset{\shortmid\shortmid}{=}O_2\overset{\shortmid\shortmid}{=}\overset{\shortmid\shortmid}{C}O$

Diallogite or manganous " $\quad Mn\overset{\shortmid\shortmid}{=}O_2\overset{\shortmid\shortmid}{=}\overset{\shortmid\shortmid}{C}O$

Smithsonite or zincic " $\quad Zn\overset{\shortmid\shortmid}{=}O_2\overset{\shortmid\shortmid}{=}\overset{\shortmid\shortmid}{C}O$

The most cursory examination of these symbols will show that they differ from each other only in the fact that one metallic atom has been replaced by another. It is not, however, every metallic atom which can thus be put in without altering the form. This is a peculiarity that is confined to certain groups of elements, which for this reason are called groups of isomorphous elements. Moreover, as a rule, there is a close resemblance between the members of any one of these groups in all their other chemical relations. These facts, like those of the last section, tend to show that the molecules of every substance have a determinate structure, which admits of a limited substitution of parts without undergoing essential change, but which is either destroyed or takes a new shape when in place of one of its constituents we force in an unconformable element. A well-known class of artificial salts, called the alums, affords even a more striking illustration of the principles of isomorphism than the simpler example we have chosen; but all the bearings of the subject cannot be understood without a knowledge of crystallography, and we must therefore refer for further details to works on mineralogy.

33. *Rational Symbols*. — Chemical formulæ, like those of the last few sections, which endeavor, by grouping together the elementary symbols, to illustrate certain classes of reactions, and to illustrate the manner in which a complex molecule may break up, are called *rational symbols*, and are to be distinguished from the simpler symbols used earlier in the book, which express only the relative proportions in which the elements are combined, and which, since they are simply expressions of the results of analysis on a concerted plan, are called *empirical symbols*. Whether these rational symbols can be regarded in any sense as indicating the actual grouping of the material atoms is very doubtful, although facts like those stated above would seem to indicate that such may be the case, at least to a limited extent. It is difficult, for example, to resist the conclusion that in alcohol and its congeners the atoms C_2H_5 are grouped together in some sense apart from the rest of the molecule; but then we have no evidence of this grouping apart from the reactions of these compounds, and, until greater certainty is reached, it is not best to attach a significance to our symbols beyond the truths they are known to illustrate.

It is objected to the use of rational symbols that they bias the judgment on the side of some theory, of which they are more or less the exponents. But when they are used in the sense stated above, this objection has no force, for the reactions they prefigure are no less facts than the definite proportions they conventionally represent, and we employ one mode of grouping the symbols or another, as will best indicate the reactions we are studying. Moreover, as science advances, we have every reason to believe that we shall gain more and more knowledge of the actual relations between the parts of a material molecule, and as has already been intimated, there can hardly be a doubt that in some cases our rational symbols do express even now actual knowledge of this sort, however crude and partial it may be. Our present typical symbols are indeed the expressions of partial generalizations, which, however imperfect, have an element of truth. Hence it is that they have pointed out new lines of investigation, have led to new discoveries, and have been of the greatest value to science. They will doubtless soon be superseded by other rational symbols, expressing other partial generalizations, to serve the same purpose in their turn and be likewise forgotten. We must not, however, despise these temporary expedients of science. They are not only useful, but necessary, and cannot mislead the student if he remembers that all such aids are merely the scaffoldings around the science, on which the builders work. It is from this point of view alone that we are to look at the whole idea of chemical atoms, which lies at the basis of our modern chemical philosophy. That this idea is actually realized in the concrete form which it takes in some minds, can hardly be believed. The true chemical idea of the *atom* is more nearly represented by the corresponding Latin word *individuum*. The atom is the chemical individual, the unit, in which the mind seeks to repose for the time the individuality of that as yet undivided substance we call an element.

34. *Graphic Symbols.* — A more graphic method of representing the relations between the atoms of a molecule than that of our ordinary rational symbols has been contrived by Kékulé, and has a similar value in aiding the conceptions, and thus facilitating the study of chemistry. In describing this system we shall speak of the possibilities of combinations of any polyad atom with monad atoms as so many centres of attraction or points of attachment, and, also, as so many affinities. Kékulé represents a monad atom, with its single centre, thus, ⊙, while the symbols (⋅ ⋅), (⋅ ⋅ ⋅), (⋅ ⋅ ⋅ ⋅), &c., represent polyad atoms of different atomicities. When the several affinities are satisfied, the points are exchanged for lines pointing in the direction of the attached atoms. Thus, the symbol represents a dyad atom with its two affinities satisfied by two monad atoms, as, for example, in a molecule of water $\overset{\text{II}}{H\text{-}O\text{-}H}$. In like manner the symbol represents a molecule of nitric anhydride $\overset{\text{V}}{N_2}\overset{\text{II}}{O_5}$, and the symbol a molecule of sulphuric anhydride $\overset{\text{VI}}{S}O_3$. Molecules like these, in which all the affinities are satisfied, are said to

CHEMICAL TYPES. 71

be *saturated* or *closed*, while the atomic group $\overset{v}{N}O_2$, represented by ⬚ has one point of attraction still open, and, therefore, acts as a monad radical. So, also, the molecular group SO_2 represented by ⬚, acts as a dyad radical.

These graphic symbols enable us to illustrate several important principles which could not readily be understood without their aid.

First. In the examples given in this section thus far, the quantivalence of a group of atoms of the same element is equal to the sum of the quantivalences of all the atoms of the group. Thus, in the molecule $\overset{v}{N_2}\overset{\text{II}}{O_5}$, the group of two pentad atoms presents ten affinities, and is saturated by the group of five dyad atoms, which presents the same number of affinities in return. So, also, in the molecule SO_3, a group of three dyad atoms just saturates the single hexad atom S. Such, however, is not necessarily the case, for it frequently happens that the similar atoms of such groups are united among themselves, and that a portion of the affinities (necessarily always an even number) are thus satisfied. For example, although C is a tetrad atom, the hydrocarbons, C_2H_6, C_2H_4, and C_2H_2, are all saturated molecules, as is shown by the following graphic symbols,

C_2H_6 C_2H_4 C_2H_2

and it is evident that in the first the two carbon atoms have been united by two, in the second by four, and in the third by six, of their eight affinities, while a corresponding number of points to which hydrogen atoms might otherwise have been attached are thus closed.

In like manner we have a well-known series of hydrocarbons, whose symbols are

$$CH_4, \; C_2H_6, \; C_3H_8, \; C_4H_{10}, \; C_5H_{12}, \; C_6H_{14}, \; \&c.,$$

the molecule of each one differing from that of the last by the group CH_2. In all these compounds the carbon atoms are united among themselves at the smallest possible number of points, as is shown, in a single case, by the following graphic symbol,

C_6H_{12}

and by constructing the graphic symbols of the other members of the series, it will be easily seen that the number of affinities thus closed is in every case equal to $2n-2$, while the number remaining open is $4n-(2n-2)=2n+2$, where n stands for the number of carbon atoms in the molecule. Hence, while the groups just mentioned form saturated molecules, the atomic groups

CH_3 C_2H_5 C_3H_7 C_4H_9 C_5H_{11} &c.,
Methyl.   Ethyl.   Propyl.   Butyl.   Amyl.

act as univalent radicals. The graphic symbol of ethyl is , and in a similar way the graphic symbols of the other radicals may be easily constructed. In like manner may be also constructed the graphic symbols of the following important compound radicals, which forms a series parallel to the first, and are all evidently dyads:—

C_2H_4 C_3H_6 C_4H_8 C_5H_{10} &c.
Ethylene.   Propylene.   Butylene.   Amylene.

Here again the graphic symbols enable us to explain a remarkable fact. These last atomic groups act not only as compound radicals, but also form the molecules of definite hydrocarbons (the first in the series being the well-known olefiant gas), and the difference in these two conditions may be represented to the eye, in the case of amylene, for example, as below:—

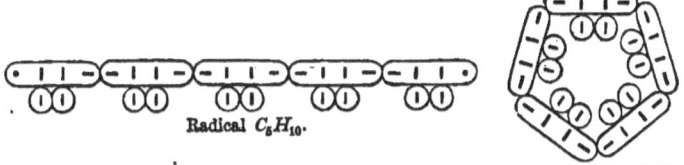

Radical C_5H_{10}.   Hydrocarbon C_5H_{10}.

The molecule in the first case is open, and presents two points of attraction, while in the second case it is closed.

CHEMICAL TYPES. 73

The members of the two classes of hydrocarbon radicals mentioned above are the characteristic constituents of an important class of compounds called alcohols, and hence they are usually called alcohol radicals. If, in these atomic groups, we substitute oxygen for a portion of the hydrogen, one atom of oxygen always taking the place of two atoms of hydrogen, we obtain still other series of radicals, which are the characteristic constituents of several important organic acids, and belong to the class of acid radicals, which will be defined in the next chapter. Among the most important of the radicals thus derived are those of the following series:—

CHO C_2H_3O C_3H_5O C_4H_7O C_5H_9O
Formyl.   Acetyl.   Propionyl.   Butyryl.   Valeryl.

and the student should construct the graphic symbol of each.

The compounds of carbon have been selected to illustrate the apparent change of atomicity which frequently accompanies the grouping together of similar atoms, because this element is peculiarly susceptible of such a mode of combination, and in fact the almost infinite variety of its compounds may be traced to this circumstance. The same phenomenon, however, is presented, although to a less marked degree, by other elements. Thus arises the remarkable fact that a group of two atoms of a bivalent element has not unfrequently only the same quantivalence as a single atom. For example, there are two compounds of mercury and chlorine $Hg=Cl_2$ represented graphically by [symbol] and $[Hg_2]=Cl_2$ represented by [symbol]. So also we have $Cu=O$ and $[Cu_2]=O_2$. We also frequently meet with another illustration of the same principle in an important class of tetrad elements whose atoms readily pair together, forming an atomic group which is sextivalent. Thus are formed the well-known compounds

$[Al_2]\equiv Cl_6$ $[Mm_2]\equiv O_3$ $[Cr_2]\equiv O_3$, &c.

When these same elements enter into combination by single atoms, they are almost invariably bivalent, and thus we have, in several cases, two very distinct classes of compounds, the one formed with the single and the other with the double atom of the element; for example,

$Fe=Cl_2$ and $[Fe_2]\equiv Cl_6$ $Fe=O$ and $[Fe_2]\equiv O_3$.

It will be noticed that although in the compounds of the second class the quantivalence of the single atoms is twice as great as it is in the first, yet their atom-fixing power is only increased by one half, and hence the name of *sesqui*-oxides or *sesqui*-chlorides, &c., which is frequently applied to them.

In order to distinguish the groups of similar atoms whose affinities are all open, from those groups where the affinities are in part closed by the union of the atoms among themselves, we may, as above, enclose the symbols of the last in brackets; and this rule will generally be followed. In most cases, however, the relations of the parts of the symbol are sufficiently evident without this aid.

Secondly. The graphic symbols illustrate another important theoretical principle, which, although almost self-evident, might be overlooked if not dwelt upon specially; namely, that on the multivalence of one or more of its atoms depends the integrity of every complex molecule. According to our present theories, no molecule can exist as an integral unit unless its parts are all bound together by such atomic clamps. Moreover, the whole virtue of a compound radical consists in the circumstance that it is an incomplete structure of the same sort, and its quantivalence is in every case equal to the number of univalent atoms (or their equivalents) which are required to complete it, or which it may be regarded as having lost. Hence the law of Wurtz finds a perfect expression in this system of graphic notation.

Thirdly. The graphic symbols illustrate most forcibly the relations of the parts of a complex molecule. Thus, for example, the symbols of alcohol and acetic acid given below show that in these compounds the dominant atom of oxygen acts as a bond uniting a complex radical to a single monad atom. They also show how it is possible that three of the atoms of hydrogen in acetic acid may stand in a very different relation to the molecule from the fourth (31). Again they show that the molecule of acetic acid differs from that of alcohol in the

Alcohol.
C_2H_5-O-H.

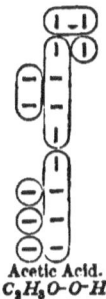

Acetic Acid.
C_2H_3O-O-H.

CHEMICAL TYPES. 75

fact that one dyad atom has taken the place of two monad atoms; and, lastly, they give form to the idea of chemical types, so far as it has any real significance. When the composition of a compound is represented in this way, all the accidental or arbitrary divisions of our ordinary notation disappear, and only those are preserved which are fundamental. We gain thus more accurate conceptions of molecular structure. We understand better the relations of the various compound radicals (compare § 28), and, above all, we thus realize the full meaning of the fundamental tenet of our new philosophy, which holds that each chemical molecule is a completed structure bound together in all its parts by a system of mutual attractions.

There is another system of graphic symbols, frequently used in works on modern chemistry, which has some advantages over the one just described. In this system the atoms are represented by small circles circumscribing the ordinary symbol, and the atomicity is indicated by dashes radiating from these circles. A few examples will sufficiently illustrate the application of this method.

$$H-O-H$$
Water
$H-O-H$

Alcohol.
C_2H_5-O-H.

Acetic Acid.
C_2H_3O-O-H.

It is obvious, however, that the circles here used are not essential, and if we omit them, and only use dashes between the dominant atoms, and also, for convenience in printing, bring the whole expression into a linear form, using commas to separate disconnected atoms, and such other signs as may be necessary to avoid ambiguity, we have at once the ordinary system of notation adopted in this book. The graphic symbols last described are merely an expansion of this system. Nevertheless, the practice of developing the ordinary symbols into either of the more graphic forms will tend to impress the full meaning of the symbols on the mind of the student, and will thus greatly aid him in acquiring a clear conception of the theory of modern chemistry.

We may, however, extend the use of dashes so as to indicate the relations of all the parts of a complex molecule by our ordinary notation. Thus we may write the symbol of alcohol

$$([C\text{-}C]\equiv H_5)\text{-}O\text{-}H,$$

or that of acetic acid

$$([C\text{-}C]\equiv H_3, O)\text{-}O\text{-}H,$$

and these expanded symbols may frequently be used to advantage in place of the graphic forms. When thus developed, the symbol indicates the quantivalence of each of the atoms of the molecule, and in every case, if the symbol is correctly written, the number of dashes will be one half of the total quantivalence of all the atoms which are thus grouped together, for each dash evidently represents two affinities.

The remarks at the close of the last section apply, of course, still more forcibly to such bold and material conceptions as these graphic symbols appear to represent, and when we recall the hooked atoms of an elder philosophy, we cannot but smile to think how closely our modern science has reproduced what we once considered as strange and grotesque fancies. But, absurd as such conceptions certainly would be, if we supposed them realized in the concrete forms which our diagrams embody, yet, when regarded as aids to the attainment of general truths, which in their essence are still incomprehensible, even these crude and mechanical ideals have the very greatest value, and cannot well be dispensed with in the study of science.

Questions and Problems.

1. To what types may the following symbols be referred, and what is the quantivalence of the different compound radicals here distinguished? Study with the same view the symbols already given in the previous chapter.

$H\text{-}(C_6H_5)$ — Benzole.
$H\text{-}(C_7H_5O)$ — Oil of Bitter Almonds.
C_2H_4 — Ethylene.
$H_2\text{=}O_2\text{=}(C_2H_2O)$ — Glycollic Acid.

$H\text{-}O\text{-}(C_6H_5)$ — Phenic Acid.
$H\text{-}O\text{-}(C_7H_5O)$ — Benzoic Acid.
$H_2\text{=}O_2\text{=}(C_2H_4)$ — Glycol.
$H_2\text{=}O_2\text{=}(C_2O_2)$ — Oxalic Acid.

$H, H, (C_6H_5)\equiv N$ — Aniline.
$H, H, (C_7H_5O)\equiv N$ — Benzamide.

$H_2, H_2, (C_2H_4)\equiv N_2$ — Ethylene diamine.
$H_2, H_2, (C_2O_2)\equiv N_2$ — Oxamide.

CHEMICAL TYPES. 77

$$H, H\text{=}N\text{-}(C_2H_2O)\text{-}O\text{-}H \qquad H, H\text{=}N\text{-}(C_2O_2)\text{-}O\text{-}H$$
Glycocol.   Oxamic Acid.

$$H, (C_7H_5O)\text{=}N\text{-}(C_2H_2O)\text{-}O\text{-}H \qquad H, H\text{=}N\text{-}(C_2O_2)\text{-}O\text{-}(C_2H_5)$$
Hippuric Acid.   Oxamethane.

2. Analyze the following reactions, and show that by comparing the reactions in each group, the typical structure of the various compounds may be inferred.

$$\underset{\text{Chlorine gas.}}{Cl\text{-}Cl} \; + \; \underset{\text{Hydrogen gas.}}{H\text{-}H} \; = \; \underset{\text{Hydrochloric Acid.}}{HCl} \; + \; \underset{\text{Hydrochloric Acid.}}{HCl}$$

$$Cl\text{-}Cl \; + \; \underset{\text{Oil of Bitter Almonds.}}{(C_7H_5O)\text{-}H} \; = \; \underset{\text{Benzoyl Chloride.}}{(C_7H_5O)\text{-}Cl} \; + \; HCl$$

$$H\text{-}Cl \; + \; \underset{\text{Potassic Hydrate.}}{K\text{-}O\text{-}H} \; = \; \underset{\text{Potassic Chloride.}}{KCl} \; + \; \underset{\text{Water.}}{H\text{-}O\text{-}H}$$

$$H\text{-}Cl \; + \; \underset{\text{Alcohol.}}{(C_2H_5)\text{-}O\text{-}H} \; = \; \underset{\text{Ethylic Chloride.}}{(C_2H_5)\text{-}Cl} \; + \; H\text{-}O\text{-}H$$

$$\underset{\text{Sulphohydric Acid.}}{H, H\text{=}S} \; + \; \underset{\text{Phosphoric Chloride.}}{P\text{≡}Cl_5} \; = \; P\text{≡}Cl_3, S \; + \; HCl \; + \; HCl$$

$$\underset{\text{Thiacetic Acid.}}{H, (C_2H_3O)\text{=}S} \; + \; P\text{≡}Cl_5 \; = \; P\text{≡}Cl_3, S \; + \; \underset{\text{Acetyl Chloride.}}{(C_2H_3O)\text{-}Cl} \; + \; HCl$$

$$\underset{\text{Potassic Hydrate.}}{K_2\text{=}O_2\text{=}H_2} + \underset{\text{Cyanic Acid.}}{(CO), H\text{≡}N} = \underset{\text{Potassic Carbonate.}}{K_2\text{=}O_2\text{=}(CO)} + \underset{\text{Ammonia.}}{H, H, H\text{≡}N}$$

$$K_2\text{=}O_2\text{=}H_2 + (CO), \underset{\text{Cyanic Ether.}}{(C_2H_5)\text{≡}N} = K_2\text{=}O_2\text{=}(CO) + H, H, \underset{\text{Ethylamine.}}{(C_2H_5)\text{≡}N}$$

3. What would be the symbols of cyanic acid and cyanic ether (see last problem), on the supposition that they contain the radical cyanogen, and are formed after the water type? Is the following reaction compatible with that last given?

$$\underset{\text{Cyanetholine.}}{K\text{=}O\text{=}H} + (C_2H_5)\text{-}O\text{-}(CN) = \underset{\text{Alcohol.}}{(C_2H_5)\text{-}O\text{-}H} + \underset{\text{Potassic Cyanate.}}{K\text{-}O\text{-}(CN).}[1]$$

and if not, what conclusion must you draw in regard to the two compounds cyanic ether and cyanetholine?

4. What bearing have the phenomena of substitution on the doctrine of chemical types? Does the circumstance that the proper-

---

[1] This product in the actual process is decomposed by the excess of potash into potassic carbonate and ammonia.

ties of the substitution products are frequently quite different from those of the original substance invalidate the doctrine?

5. How does the action of chlorine on acetic acid indicate that this compound is fashioned after a determinate type? On what particular fact does this evidence chiefly rest?

6. What bearing have the phenomena of isomorphism on the doctrine of types? Enforce the argument by some familiar illustration.

7. The radical allyl C_3H_5 is univalent in oil of garlic $(C_3H_5)_2\text{=}S$, and in allylic alcohol $(C_3H_5)\text{-}O\text{-}H$, but atrivalent in glycerine $(C_3H_5)_3\text{≡}O_3\text{≡}H_3$. Moreover, this radical when set free doubles, forming a volatile hydrocarbon oil, which has the composition $(C_3H_5)\text{≡}(C_3H_5)$, and which combines directly with bromine, the resulting product having the symbol $(C_3H_5)\text{-}(C_3H_5)\text{≡}Br_4$. Represent these symbols by the graphic method, and thus explain the different relations of the radical.

8. Represent the symbols of phenic acid and benzoic acid by the second graphic method, and explain why the radical phenyl (C_6H_5) and benzoyl (C_7H_5O) are only univalent.

9. Why is it that the addition of the atoms CH_2 does not change the atomicity of a radical?

10. What is the quantivalence of Al in the symbol $[Al\text{-}Al]\text{≡}Cl_6$? Is there any difference in the quantivalence of Fe in the two compounds $Fe\text{=}O_2\text{=}CO$ and $[Fe\text{-}Fe]\text{≡}O_6\text{≡}SO_2$? Answer the questions by the aid of graphic symbols.

11. Is there any difference in the quantivalence of nitrogen in potassic nitrite $K\text{-}O\text{-}NO$ and potassic nitrate $K\text{-}O\text{-}NO_2$?

12. Represent by graphic symbols the difference between cyanic ether and cyanetholine (see problems 2 and 3 above).

13. The symbol $[Hg_2]Cl_2$ represents a single molecule, while Na_2Cl_2 represents two molecules, and would be more properly written $2NaCl$. What is the difference in the two cases?

14. Represent by the graphic method the symbols of potassic carbonate $K_2\text{=}O_2\text{=}(CO)$ and potassic oxalate $K\text{=}O\text{=}(C_2O^2)$, and show that both form a perfect molecular unit.

15. Represent by the graphic method the following symbols;

$H_2\text{=}O_2\text{=}(C_3H_6)$ (Propyl Glycol.);

$H_2\text{=}O_2\text{=}(C_3H_4O)$ (Lactic Acid.);

$$H_2{=}O_2{=}(C_3H_2O_2) \quad \text{(Malonic Acid)};$$

$$H_2{=}O_2{=}(C_3O_3) \quad \text{(Unknown)},$$

and thus show that they are formed after the same type.

16. What is the atom-fixing power or quantivalence of the elements and radicals, which appear in the various symbols given in this chapter? Develop these symbols, and show that they represent in each case a single perfect molecule.

N. B. The student should practice developing the ordinary molecular symbols into the graphic forms described above, until he is perfectly familiar with the method, and has acquired a clear conception of the different types of molecular structure.

Q 115 - 70 bottom of top 71

Q 116 - 71 page

Q 117 71 bottom of top 72

Hydrocarbons must [illegible]

C_nH_{2n+2}

different values to n get whole set of Hydrocarbons.

n = 1

CH_4

suppose 1 of the H removed then we have [illegible]

n = 2

C_2H_6

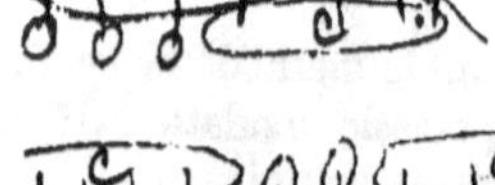

remove another atom of H and we make them [illegible]

n = 3

C_3H_8

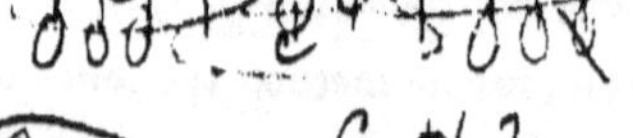

Q 118 n=1

CH_2
CH_2
C_2H_4

CH_2

Q 119

CHAPTER IX.[1]

BASES, ACIDS, AND SALTS.

35. *Hydrates, Alkalies, Bases.* — It is not unfrequently the case that the technical terms of a science remain in use long after they have lost their original meaning. This is peculiarly true of those which we have placed at the head of this section. They have, with the exception of the first, come down to us from the period of alchemy, and are still retained in the language of trade and in many works on practical science, with a peculiar meaning which they have acquired during the last hundred years under the teaching of the dualistic theory. Since they, in many cases at least, suggest erroneous conceptions in regard to the constitution of chemical compounds, it would be well if they could be discarded altogether; but, as this is impracticable, we must endeavor to give to them as definite a meaning as possible.

The term "hydrate" is applied to a class of compounds which were formerly supposed to contain water as such, but which are now believed to have no closer relation to water than is indicated by the circumstance that they have the same type, and may be formed from water by replacing one of its hydrogen atoms with some metal. Thus, by acting on water with potassium, we obtain potassic hydrate; or, if we use sodium, we obtain sodic hydrate.

$$2H\text{-}O\text{-}H + K\text{-}K = 2K\text{-}O\text{-}H + H\text{-}H$$
<small>Water.　　Potassium.　Potassic Hydrate.　Hydrogen Gas.</small>

$$2H\text{-}O\text{-}H + Na\text{-}Na = 2\,Na\text{-}O\text{-}H + H\,H$$
<small>Water.　　Sodium.　　Sodic Hydrate.　Hydrogen Gas.</small>

[42]

Both of these hydrates, and also those of the very rare but closely allied metals, lithium, cæsium, and rubidium, are very

---

[1] In studying this chapter the student should endeavor to remember the names and symbols of the different compounds mentioned. Hitherto we have been chiefly employed with the forms of the symbols, and this exercise of the memory has not been expected.

soluble in water, and yield solutions which corrode the skin, and convert the fats into soaps. To all the substances known to them which possessed these caustic qualities the alchemists gave the name of *alkalies*, and this term is now applied to the five hydrates just enumerated. The first two of these are commercial products, and have important applications in the arts. They all differ from the hydrates of other metals in that they cannot be decomposed by heat alone.

Again, if we act on water with calcium or magnesium, we obtain calcic or magnesic hydrate; but the double atom of water is then decomposed by these bivalent metals.

$$\underset{\text{Water.}}{H_2\text{=}O_2\text{=}H_2} + \underset{\text{Calcium.}}{\overset{\text{II}}{Ca}} = \underset{\text{Calcic Hydrate.}}{Ca\text{=}O_2\text{=}H_2} + \underset{\text{Hydrogen Gas.}}{H\text{-}H} \qquad [43]$$

$$\underset{\text{Water.}}{H_2\text{=}O_2\text{=}H_2} + \underset{\text{Magnesium.}}{\overset{\text{II}}{Mg}} = \underset{\text{Magnesic Hydrate.}}{Mg\text{=}O_2\text{=}H_2} + \underset{\text{Hydrogen Gas.}}{H\text{-}H}$$

These two hydrates, as well as those of the allied metals, barium and strontium, although much less soluble in water than the alkalies, still dissolve in this common solvent to a limited extent, and manifest decided caustic qualities. When dry they have an earthy appearance, and hence are frequently known as the alkaline earths. They also differ from the true alkalies in the fact that they are readily decomposed by heat; and since they are then resolved into water and a metallic oxide, as the following reaction shows, the opinion formerly entertained in regard to their composition was not unnatural.

$$Mg\text{=}O_2\text{=}H_2 \underset{\text{When heated.}}{=} MgO + H_2O \qquad [44]$$

Moreover, when the anhydrous oxides are mixed with water, they enter into direct union with a portion of the liquid. This combination is usually attended with the evolution of great heat, and the process is known as slaking.

$$CaO + H_2O = Ca\text{=}O_2\text{=}H_2. \qquad [45]$$

There are many other metallic hydrates which are still more readily decomposed by heat. These, as a rule, cannot be formed by the direct union of the corresponding metallic oxide and water, but may be obtained by adding to a solution of a salt of the metal one of the soluble hydrates mentioned above. Thus,—

$$(\underset{\text{Cupric Chloride.}}{CuCl_2} + 2Na\text{-}O\text{-}H + Aq) = (\underset{\text{Cupric Hydrate.}}{Cu\text{=}O_2\text{=}H_2} + \underset{\text{Sodic Chloride.}}{2NaCl} + Aq)$$

[46]

$$(\underset{\text{Zincic Chloride.}}{ZnCl_2} + 2K\text{-}O\text{-}H + Aq) = (\underset{\text{Zincic Hydrate.}}{Zn\text{=}O_2\text{=}H_2} + \underset{\text{Potassic Chloride.}}{2KCl} + Aq)$$

$$(\underset{\text{Ferric Chloride.}}{[Fe_2]Cl_6} + 3Ba\text{=}O_2\text{=}H_2 + Aq) = (\underset{\text{Ferric Hydrate.}}{[Fe_2]\equiv O_3\equiv H_6} + \underset{\text{Baric Chloride.}}{3BaCl_2} + Aq)$$

The hydrates are regarded by some chemists as compounds of the metal with the compound radical hydroxyl, and their symbols are then written after a simpler type, thus,—

$$\underset{\text{Calcic Hydrate.}}{Ca\text{=}(HO)_2} \qquad \underset{\text{Ferrous Hydrate.}}{Fe\text{=}(HO)_2} \qquad \underset{\text{Chromic Hydrate.}}{[\overset{VI}{Cr_2}]\equiv(HO)_6}$$

Ammonia. — Closely allied to these metallic hydrates is a very remarkable compound, formed by dissolving ammonia gas, NH_3, in water. Although the product resembles, in many of its physical relations, a simple solution of gas in water, yet the compound in all its chemical relations acts like a metallic hydrate,

$$\underset{\text{Ammonia Gas.}}{NH_3} + \underset{\text{Water.}}{H_2O} = \underset{\text{Ammonic Hydrate.}}{\overset{I}{NH_4}\text{-}O\text{-}H}$$

which has led chemists to write its symbol after the type of water, and to assume the existence of a univalent compound radical $\overset{I}{NH_4}$, to which has been given the name of ammonium.

Metallic Oxides or Basic Anhydrides. — Closely allied to the metallic hydrates, in the relation we are now considering, are many of the simple compounds of the metals with oxygen which are called in general metallic oxides. Such compounds as

$$\underset{\text{Calcic Oxide.}}{Ca\text{=}O} \quad \underset{\text{Baric Oxide.}}{Ba\text{=}O} \quad \underset{\text{Plumbic Oxide.}}{Pb\text{=}O} \quad \underset{\text{Ferrous Oxide.}}{Fe\text{=}O} \quad \underset{\text{Cupric Oxide.}}{Cu\text{=}O} \quad \underset{\text{Argentic Oxide.}}{Ag_2\text{=}O}$$

may be regarded as formed from one or more molecules of water, by replacing all the atoms of hydrogen with those of some metal; and these oxides as well as the hydrates before mentioned are frequently classed together under the common title of *bases*, although it would be best to confine this term to the metallic hydrates alone, and to distinguish the *basic oxides* as *basic anhydrides*. (37)

Salts. — The atoms of hydrogen still remaining in a metallic hydrate may be replaced with the atoms of a well-defined class of non-metallic elements and compound radicals; and, for a reason which will soon appear, the replacing atoms are called acid or negative radicals.[1]

From this replacement results a new class of compounds we call *salts*. Thus, —

$K\text{-}O\text{-}H$ gives $K\text{-}O\text{-}Cl$, also $K\text{-}O\text{-}NO_2$ and $K\text{-}O\text{-}(C_2H_3O)$
Potassic Hydrate.   Potassic Hypochlorite.   Potassic Nitrate.   Potassic Acetate.

$Ca\text{=}O_2\text{=}H_2$ gives $Ca\text{=}O_2\text{=}SO_2$ $Ca\text{=}O_2\text{=}CO$ $Ca\text{=}O_2\text{=}(C_2H_3O)_2$
Calcic Hydrate.   Calcic Sulphate.   Calcic Carbonate.   Calcic Acetate.

36. *Acids.* — Opposed in chemical properties to the so-called bases is another very important class of compounds called *acids*. They derive their name from the fact that they are generally soluble in water and have a sharp or sour taste, although there are many exceptions to the rule. Like the bases, they all contain hydrogen; but this hydrogen can no longer be replaced by non-metallic elements or negative radicals, but only by metallic elements and positive radicals, and it is herein that the chief distinction lies. Moreover, the opposition of these two classes of compounds also appears in the fact that, while in bases the replaceable hydrogen atoms are united to a metallic atom or positive radical, which for this reason we frequently call a basic radical, in the acids, on the other hand,

---

[1] The word radical, as used in chemistry, stands for any atom or group of atoms, which is, for the moment, regarded as the principal constituent of the molecules of a given compound, and which does not lose its integrity in the ordinary chemical reactions to which the substance is liable. The distinction between basic and acid radicals (or positive and negative radicals as they are more frequently called) will become clear as we advance. It is sufficient for the present to state that, although these terms imply an *opposition of relations* rather than a difference of qualities, yet, as a general rule, the metallic atoms are basic radicals, while the non-metallic atoms are acid radicals. Moreover it may be added, that among compound radicals those consisting of carbon and hydrogen alone are usually basic, and those containing also oxygen usually acid; and, further, that of the two most important radicals containing nitrogen, ammonium (NH_4) is strongly basic, and cyanogen (CN) as decidedly acid. In this book, with few exceptions, the basic radicals are always placed on the left-hand, and the acid radicals on the right-hand side, of the molecular symbols.

these same hydrogen atoms are united as a rule to a nonmetallic atom or negative radical, frequently, also, called as above an acid radical. In most cases there is a vinculum which unites the two parts of the molecule; and both in acids and in bases this vinculum consists usually of one or more oxygen atoms, although in a large class of acids the hydrogen atoms are united directly to the radical without any such connection. The acids of this class have by far the simplest constitution; and we will give examples of these first, adding in each case a reaction to illustrate the acid relations of the compound. In studying these reactions, it must be borne in mind that the evidence of acidity is in each case to be found in the fact that one or more of the hydrogen atoms of the compound may be replaced by positive radicals or metallic atoms. This replacement may be obtained in one of four ways, — by acting on the acid, either with the metal itself, or with a metallic oxide, or with a metallic base, or with a metallic salt.

$$(2HCl + Aq) + NaNa = (2\ NaCl + Aq) + \underline{H\text{-}H}$$
Hydrochloric Acid.   Sodium.   Sodic Chloride.

$$(2HCl + Aq) + ZnO = (ZnCl_2 + H_2O + Aq)$$
Zincic Oxide.   Zincic Chloride.

[47]

$$(HBr + K\text{-}O\text{-}H + Aq) = (KBr + H_2O + Aq)$$
Hydrobromic Acid.   Potassic Hydrate.   Potassic Bromide.

$$(HI + Ag\text{-}O\text{-}NO_2 + Aq) = AgI + (H\text{-}O\text{-}NO_2 + Aq)$$
Hydriodic Acid.   Argentic Nitrate.   Argentic Iodide.   Nitric Acid.

We will next give examples of more complex acids, in which the two parts of the molecule are united by a vinculum of oxygen atoms.

$$(H\text{-}O\text{-}(C_2H_3O) + Na\text{-}O\text{-}H + Aq) = (Na\text{-}O\text{-}(C_2H_3O) + H_2O + Aq)$$
Acetic Acid.   Sodic Hydrate.   Sodic Acetate.

$$(H_2\text{=}O_2\text{=}SO_2 + Aq) + CuO = (Cu\text{=}O_2\text{=}SO_2 + H_2O + Aq)$$
Sulphuric Acid.   Cupric Oxide.   Cupric Sulphate.

$$(H_3\text{=}O_3\text{=}PO + 3K\text{-}O\text{-}H + Aq) = (K_3\text{=}O_3\text{=}PO + 3H_2O + Aq)$$
Phosphoric Acid.   Potassic Hydrate.   Potassic Phosphate.

Such acids as these are called oxygen acids. Like the hydrates, they may be regarded as compounds of hydroxyl, but with negative instead of positive radicals, thus: —

BASES, ACIDS, AND SALTS. 85

$HO\text{-}NO_2$ $(HO)_2\!=\!SO_2$ $(HO)_3\!\equiv\!PO.$
Nitric Acid.   Sulphuric Acid.   Phosphoric Acid.

This mode of writing the symbols is not only frequently convenient, but has been of real value by bringing out unexpected and important relations. It does not, however, indicate any fundamental difference of opinion in regard to the constitution of these hydrates, and this at once appears when the symbols are put into the graphic form.

When an acid, like acetic acid, contains but one atom of hydrogen, which is replaceable by a metallic atom or a positive radical, it is called monobasic; when, like sulphuric acid, it contains two such hydrogen atoms, it is called dibasic; when, like phosphoric acid, it contains three, it is tribasic, &c. Moreover, one evidence of this difference of basicity is found in the fact that whereas a monobasic acid can only form one salt with a univalent radical, a bibasic acid can form two, and a tribasic three. Thus, while we have only one sodic nitrate, there are two sodic sulphates and three sodic phosphates.

$Na\text{-}O\text{-}NO_2$ $H_2Na\!\equiv\!O_3\!\equiv\!PO$
Sodic Nitrate.   Acid Sodic Phosphate.

$H,Na\!=\!O_2\!=\!SO_2$ $HNa_2\!\equiv\!O_3\!\equiv\!PO$
Acid Sodic Sulphate.   Neutral Sodic Phosphate.   [48]

$Na_2\!=\!O_2\!=\!SO_2$ $Na_3\!\equiv\!O_3\!\equiv\!PO$
Neutral Sodic Sulphate.   Basic Sodic Phosphate.

There is, however, but one calcic sulphate, for, since the calcium atoms are bivalent, a single one is sufficient to replace both of the hydrogen atoms in the acid.

37. *Acid Anhydrides.* — Besides the acids properly so called, all of which contain hydrogen, there is another class of compounds which bear the same relation to the true acids which the metallic oxides bear to the true bases. To avoid confusion, compounds of this class have been distinguished as *anhydrides*,[1] and they may be regarded as one or more molecules of water in which all the hydrogen has been replaced by negative or acid radicals. As among the most important of these we may enumerate Sulphuric Anhydride $SO_2\!=\!O$ or SO_3, Nitric Anhy-

---

[1] More precisely *acid anhydrides*, but as the basic anhydrides are usually called simply metallic oxides, the qualifying term is seldom added.

dride $(NO_2)_2{=}O$ or N_2O_5, Carbonic Anhydride $CO{=}O$ or CO_2, Phosphoric Anhydride $(\overset{\text{I}}{PO_2})_2{=}O$ or P_2O_5, and Silicic Anhydride $\overset{\text{IV}}{Si}{\equiv}O_2$. Most of the anhydrides unite directly with water to form acids, and several of the acids, when heated, give off water and are resolved into anhydrides. [Compare 44 and 45.]

$$H_2O + SO_3 = H_2{=}O_2{=}SO_2$$
$$3H_2O + P_2O_5 = 2H_3{\equiv}O_3{\equiv}PO$$
$$\underset{\text{Silicic Acid.}}{H_4{\equiv}O_4{\equiv}Si} = \underset{\text{Silicic Anhydride.}}{SiO_2} + 4H_2O$$
$$\underset{\text{Boric Acid.}}{2H_3{\equiv}O_3{\equiv}B} = \underset{\text{Boric Anhydride.}}{B_2O_3} + 3H_2O$$

[49]

Moreover in many cases these anhydrides will combine directly with the metallic oxides to form salts; and the reactions are best indicated by a rational formula, which represents the oxide and anhydride as radicals in the resulting compound. Thus, baric oxide burns in the vapor of sulphuric anhydride, yielding baric sulphate; and lime also unites directly with the same anhydride, although with less energy, forming calcic sulphate.

$BaO + SO_3 = BaO, SO_3$ and $CaO + SO_3 = CaO, SO_3$

We are thus led to the old formulæ of the dualistic system, according to which the metallic oxides were the only true bases, the anhydrides were the only true acids, and the two were regarded as paired in all true salts. But, although in its modern theories our science has fortunately left the ruts to which the dualistic ideas for so long limited its progress, yet it must be remembered, that, according to our present definitions, these dualistic formulæ are perfectly legitimate, and still give the simplest exposition of a large number of important facts.

38. *Salts.*— The definition of the term "salt" has been clearly implied in the definitions of "base" and "acid" already given. It is any acid in which one or more atoms of hydrogen have been replaced with metallic atoms or basic radicals; it is any base in which the hydrogen atoms have been more or less replaced by non-metallic atoms or acid radicals; or it may be the

product of the direct union of a metallic oxide and an anhydride. A neutral salt is, properly speaking, one in which all the hydrogen atoms, whether of base or acid, have been replaced as just stated. A basic salt is one in which one or more of the hydrogen atoms of the base remain undisturbed, and therefore still capable of replacement by acid radicals. An acid salt is one in which one or more of the hydrogen atoms of the acid remain undisturbed, and therefore capable of replacement by basic radicals.

But, besides the basic and acid salts, which come under these definitions, there are also others which can be most simply defined as consisting of several atoms of the metallic oxide to one of anhydride, or of several atoms of anhydride to one of the metallic oxide.

As an example of acid salts of the second class we have, besides the two sodic sulphates mentioned on page 61, also a third, which may be written $Na_2O, 2SO_3$. This is easily obtained by simply heating the acid sulphate.

$$2(H, Na^=O_2^=SO_2) = Na_2O, 2SO_3 + H_2O \quad [50]$$
$$\text{Acid Sodic Sulphate.} \qquad \text{Sodic Bisulphate.} \quad \text{Water.}$$

If heated to a still higher temperature, one atom of the anhydride is set free, and the salt falls back into the neutral sulphate.

$$Na_2O, 2SO_3 = Na_2O, SO_3 + SO_3$$
$$\text{Bisulphate.} \qquad \text{Neutral Sulphate.} \qquad \text{Anhydride.}$$

This reaction justifies the dualistic form given to the symbol; but other relations of the bisulphate may be better expressed by the following typical formula, —

$$Na_2^=O_2^=(SO_2\text{-}O\text{-}\overset{\text{II}}{S}O_2) = Na_2^=O_2^=SO_2 + SO_3$$
$$\text{Sodic Bisulphate.} \qquad \text{Neutral Sulphate.} \quad \text{Anhydride.}$$

in which a group of two atoms of SO_2, soldered together by one atom of oxygen, acts as a bivalent radical.

As an example of a basic salt of the second class we have, in addition to the two plumbic acetates of the normal type,

$$Pb^=O_2^=(C_2H_3O)_2 \quad \text{and} \quad Pb^=O_2^=(C_2H_3O), H$$
$$\text{Neutral Plumbic Acetate.} \qquad\qquad \text{Basic Plumbic Acetate.}$$

a third salt containing three times as much lead, —

$$(Pb\text{-}O\text{-}Pb\text{-}O\text{-}\overset{\text{II}}{Pb}) = O_2 = (C_2H_3O)_2, \qquad [51]$$
Triplumbic Acetate.

in which a group of three atoms of lead, soldered together by two atoms of oxygen, acts as a bivalent radical. It is evident that, theoretically, any number of multivalent radicals might be united in this way, and also that the complex radical thus formed will have a quantivalence easily determined by estimating the number of bonds which remain unsatisfied; but, practically, the grouping cannot be carried to a very great extent, for the stability of the radical diminishes with its complexity, and a condition is soon reached when it can no longer sustain, if we may so express it, its own weight. Moreover, while some radicals, like the atoms of lead, copper, mercury, and iron, are prone to group themselves in this way, the larger number show but little tendency to this mode of union.

The symbols of these acetates may also be written on the dualistic type, which represents them as compounds of plumbic oxide, PbO, and acetic anhydride, $C_4H_6O_3$. We have, then, —

$$PbO, C_4H_6O_3 \quad \text{and} \quad 3PbO, C_4H_6O_3 \qquad [52]$$
Neutral Plumbic Acetate.   Triplumbic Acetate.

and we may thus best illustrate the important fact that the second compound is prepared by combining with the first an additional quantity of plumbic oxide.

It will appear on reviewing the symbols of the acids, bases, and salts given in this section, that, in by far the greater number, the two parts of the molecule are held together by one or more atoms of oxygen, which act as a vinculum. Such compounds are called oxygen salts, using the word salt, as is frequently done, to stand for acids and bases, as well as for the true metallic salts; and in fact they all belong to the same type of chemical compounds. Since oxygen plays so important a part in terrestrial nature, we might well expect that these oxygen compounds would hold a very conspicuous place in our chemical science, — and such is indeed the fact. During the dualistic period the study of chemistry was almost wholly confined to the oxygen compounds, and, even now, they occupy by far the largest share of a chemist's attention.

There is, however, another element, namely, sulphur, which is capable of filling the place occupied by oxygen in many of its compounds, and thus may be formed a distinct class of bodies which are called sulphur salts. These compounds are not nearly so numerous as the oxygen salts, and have not been so well studied, so that a few examples will be sufficient to illustrate their general composition, and the relations which they bear to the corresponding oxygen compounds.

Oxygen Salts.	Sulphur Salts.
$H\text{-}O\text{-}H$	$H\text{-}S\text{-}H$
Water or Hydric Acid.	Sulphohydric Acid.
$K\text{-}O\text{-}H$	$K\text{-}S\text{-}H$
Potassic Hydrate.	Potassic Sulphohydrate.
$K_2\text{=}O_2\text{=}CO$	$K_2\text{=}S_2\text{=}CS$
Potassic Carbonate.	Potassic Sulphocarbonate.

39. *Test-Papers.* — The soluble bases and acids, when dissolved in water, cause a striking change of color in certain vegetable dyes, and these characteristic reactions give to the chemist a ready means of distinguishing between these two important classes of compounds. The two dyes chiefly used for this purpose are turmeric and litmus, and strips of paper colored with the dyes are employed in testing. Turmeric paper, which is naturally yellow, is turned brownish red by bases, while litmus paper, which is naturally blue, is turned red by acids, and in both cases the natural color is restored by a compound of the opposite class.

If to a solution of a strong base, like sodic hydrate, we add slowly and carefully a solution of a strong acid, like sulphuric, we shall at last reach a condition in which the solution affects neither test-paper, and it is then said to be *neutral*. On evaporating this solution we obtain a neutral salt, like sodic sulphate, and the presence in the solution of the slightest excess of acid or base beyond the amount required to form this salt would have been made evident by the test-papers. In such cases, we may therefore use these test-papers to distinguish between acid, basic, and neutral salts, but only with great caution; for we find that when, as in acid-carbonate of soda, a strong base is associated with a weak acid, the reaction is still basic, although the acid may be greatly in excess, and, on the other hand, when, as in cupric sulphate, a weak base has been associated with a strong acid, the reaction may be strongly acid even in the basic salts. The explanation of these apparent anomalies is to be found in the fact that these colored reagents are all salts themselves, and the reactions examples of metathesis. The coloring matter of these dyes is an acid which varies its tint according as the hydrogen atoms have or have not been replaced; and when, for any reason, the acid or base of the salt examined is not in a condition to determine the necessary metathesis, the characteristic change of color does not take place.

Unfortunately, the facts just stated have led to great confusion in the use of the words "acid" and "basic" as applied to salts, since these terms sometimes have reference solely to the number of atoms of hydrogen, in the acid or base, which have not been replaced in the formation of the salt, and at other times refer to the reactions of the salt on the colored reagents just described. A confusion of this sort must have been noticed in the names of the three phosphates of soda on page 85. The so called neutral phosphate is theoretically an acid salt, and the basic phosphate a neutral salt, but the salts give with test-papers the reactions which their names indicate. The theoretical is the only legitimate use, and the one we shall adhere to in this book, except in regard to names of compounds which cannot be arbitrarily changed.

40. *Alcohols, Fat Acids, Ethers.* — The hydrocarbon radicals mentioned in § 34 yield a very large number of compounds after the type of water, which are closely allied to the hydrates and anhydrides, both acid and basic, just described. If one of the hydrogen atoms in the molecule of water is replaced by either of the univalent basic radicals, methyl, ethyl, propyl, &c., we obtain a class of compounds, called alcohols, of which our common alcohol is the most important. On the other hand, if the atom of hydrogen is replaced by one of the univalent acid radicals, formyl, acetyl, propionyl, &c., we obtain an important class of acid compounds, of which acetic acid (vinegar) is the best known, but which also includes a large number of fatty substances closely related to our ordinary fats. Hence the name Fat Acids, by which this class of compounds is generally designated.

BASES, ACIDS, AND SALTS. 91

Basic Hydrates or Alcohols.
Methylic Alcohol (wood spirits) CH_3-O-H.
Ethylic Alcohol (common alcohol) C_2H_5-O-H.
Propylic Alcohol C_3H_7-O-H.
Butylic Alcohol C_4H_9-O-H.
Amylic Alcohol (fusel oil) C_5H_{11}-O-H.
(With six others already known.)

Acid Hydrates, Fat Acids.
Formic Acid H-O-CHO.
Acetic Acid H-O-C_2H_3O.
Propionic Acid H-O-C_3H_5O.
Butyric Acid H-O-C_4H_7O.
Valerianic Acid H-O-C_5H_9O.
(With fifteen others already known.)

If now we replace both of the hydrogen atoms of water by the same basic radicals mentioned above, we obtain a class of compounds called ethers, which correspond to the metallic oxides or basic anhydrides; and if we replace the two hydrogen atoms by the corresponding acid radicals, we obtain a similar series of acid anhydrides. Lastly, if we replace one of the hydrogen atoms by a basic radical, and the other by an acid radical, we get a class of compounds also called ethers (but distinguished as compound ethers), which correspond to the salts.

Examples of Anhydrides.

1. Simple Ethers.

Methylic Ether CH_3-O-CH_3 or $(CH_3)_2$=O.
Ethylic Ether (common ether) C_2H_5-O-C_2H_5 or $(C_2H_5)_2$=O.

2. Mixed Ethers.

Methyl-ethyl Ether CH_3-O-C_2H_5.
Ethyl-amyl Ether C_2H_5-O-C_5H_{11}.

3. Compound Ethers.

Acetic Ether C_2H_5-O-C_2H_3O.
Butyric-methyl Ether CH_3-O-C_4H_7O.

4. Acid Anhydrides.

Acetic Anhydride C_2H_3O-O-C_2H_3O or $(C_2H_3O)_2$=O.
Valerianic Anhydride C_5H_9O-O-C_5H_9O or $(C_5H_9O)_2$=O.

The positive radicals, of which the alcohols consist, hold an intermediate position between the strong basic radicals on the one hand, and the strong acid radicals on the other, and the same is true of the alcohols themselves, which hold a middle place between the strong basic and the strong acid hydrates. This is indicated by the following reactions; in what way it is left to the student to inquire.

$$2H\text{-}O\text{-}C_2H_5 + K\text{-}K = 2K\text{-}O\text{-}C_2H_5 + H\text{-}H$$

$$2\ CH_3\text{-}O\text{-}H + H_2\text{=}O_2\text{=}SO_2 = (CH_3)_2\text{=}O_2\text{=}SO_2 + 2\ H_2O$$

41. Glycols.—The class of hydrates described in the last section belong to the simple type of water. But we have also a class of analogous compounds belonging to the type of water doubly condensed. If in the double molecule of water ($H_2\text{=}O_2\text{=}H_2$) we replace one of the pairs of hydrogen atoms by either of the bivalent positive radicals, ethylene, propylene, butylene, &c., we obtain a series of compounds closely resembling the alcohols, called glycols, and by substituting the related negative radicals we obtain two series of acid hydrates, which stand in the same relation to the glycols that the fat acids bear to the alcohols. These relations are shown in the following scheme, which, however, includes only the five first members of each of these three series of compounds. It should be noticed in this connection that each of the *bivalent positive* radicals yields *two* related *negative* radicals, while the *univalent positive* radicals of the last section yield only one such negative radical; and moreover that the acids in the first series, although diatomic, are only monobasic, while those in the second series are both diatomic and bibasic (43).

$C_2H_4\text{=}O_2\text{=}H_2$ Ethylic Glycol.	$H_2\text{=}O_2\text{=}C_2H_2O$ Glycolic Acid.	$H_2\text{=}O_2\text{=}C_2O_2$ Oxalic Acid.
$C_3H_6\text{=}O_2\text{=}H_2$ Propylic Glycol.	$H_2\text{=}O_2\text{=}C_3H_4O$ Lactic Acid.	$H_2\text{=}O_2\text{=}C_3H_2O_2$ Malonic Acid.
$C_4H_8\text{=}O_2\text{=}H_2$ Butylic Glycol.	$H_2\text{=}O_2\text{=}C_4H_6O$ Acetonic Acid.	$H_2\text{=}O_2\text{=}C_4H_4O_2$ Succinic Acid.
$C_5H_{10}\text{=}O_2\text{=}H_2$ Amylic Glycol.	$H_2\text{=}O_2\text{=}C_5H_8O$	$H_2\text{=}O_2\text{=}C_5H_6O_2$ Lipic Acid.
$C_6H_{12}\text{=}O_2\text{=}H_2$ Hexyl Glycol.	$H_2\text{=}O_2\text{=}C_6H_{10}O$ Leucic Acid.	$H_2\text{=}O_2\text{=}C_6H_8O_2$ Adipic Acid.

Corresponding to these basic and acid hydrates we have also been able to obtain in several cases the basic and acid anhydrides, besides a very large number of compound ethers.

42. *Glycerines and Sugars.* — In the alcohols one hydrogen atom from the original typical molecule (*typical hydrogen*) remains undisturbed. In the glycols there are two such hydrogen atoms, and hence these compounds are frequently called diatomic alcohols. Our common glycerine is a triatomic alcohol, and may be regarded as formed from a molecule of water trebly condensed ($H_3^{\equiv}O_3^{\equiv}H_3$), by replacing one of the groups of hydrogen atoms with the trivalent radical glyceryl (C_2H_5). It is probable that a large number of triatomic alcohols or glycerines may hereafter be obtained, but only two are now known.

Propylic Glycerine (common glycerine) $H_3^{\equiv}O_3^{\equiv}C_3H_5.$
Amylic Glycerine $H_3^{\equiv}O_3^{\equiv}C_5H_9.$

From the glycerines we may derive acids, anhydrides, and compound ethers, bearing to each other the same relations as those derived from the alcohols of a lower order, but only a few of the possible compounds which our theory would foresee are yet known. Lastly, it appears probable that our common sugars are also constituted after the type of water greatly condensed, and are simply alcohols of a very high order of atomicity.

43. *Atomicity and Basicity of an Acid.* — By the *atomicity of a compound* is meant the number of hydrogen atoms which it retains from the original typical molecule still unreplaced, and the use of this term with reference to the *basic hydrates* has been already abundantly illustrated in this chapter. In the case of the acids a distinction must be made between atomicity and basicity, which is frequently important.

The formula of every acid may be written on the type of one or more atoms of hydrochloric acid, as $H_n R^n$, in which H_n stands for the replaceable atoms of hydrogen, and R^n for all the rest of the atoms of the molecule, which may be regarded as forming a radical with an atomicity equal to the number of replaceable hydrogen atoms. The symbols $H\text{-}\overset{\text{I}}{N}O_3$ $H_2\overset{\text{II}}{S}O_4$ $H_3^{\equiv}\overset{\text{III}}{P}O_4$ are written on this principle. In each case the acid is said to have the atomicity of the radical. The basicity of the acid, on the other hand, depends, not on the *total* number of replaceable hydrogen atoms, but on the number which may be replaced by *metallic* atoms or *basic* radicals. As a general rule, it is true that the basicity is the same as the atomicity, but this is not always the case. Thus lactic acid is diatomic but monobasic, and the same is true of the other acids homologous with it (page 92).

$$\overset{+}{H},\ \overset{-}{H}{=}(C_3H_4\overset{II}{O_3})\quad Na,\ H{=}(C_3H_4\overset{II}{O_3})\quad Na,\ (C_7H_5\overset{I}{O}){=}(C_3H_4\overset{II}{O_3})$$
Lactic Acid.   Sodic Lactate.   Sodic Benzolactate.

$$K,\ C_2H_5{=}(C_3H_4O_3)\quad\quad C_2H_5,\ C_2H_5{=}(C_3H_4O_3)$$
Potassic Ethyl-lactate.   Diethylic-lactate.

Only one atom of hydrogen can be replaced by a metallic radical, but a second may be replaced by either a negative or an alcoholic radical, as in the last three symbols, and in designating the atoms, thus differently related to the molecular structure, it is usual to call the first basic and the other alcoholic hydrogen.

We might, in like manner, distinguish between the atomicity and the *acidity* of a base, but this distinction has not been found as yet to be of practical importance.

44. *Water of Crystallization.* — Among the most striking characteristics of the class of compounds we call salts is their solubility in water and their tendency on separating from it, in consequence of either the evaporation or the cooling of the fluid, to assume definite crystalline forms. These crystals, as a general rule, are complex crystalline aggregates of molecules of the salt and molecules of water. The water is held in combination by a comparatively feeble force, and may be generally driven off by exposing the salt to the temperature of 100° C., when the crystals fall to powder. Sometimes it escapes at the ordinary temperature of the air, when the crystals, as before, fall to powder and are said to effloresce. It thus evidently appears that the water, although an essential part of the crystalline structure, is not inherent in the chemical molecule, and hence the name Water of Crystallization. The presence of

water of crystallization in a salt is expressed by writing after the symbol of the salt, and separated from it by a period, the number of molecules of water with which each salt molecule is associated. Thus we have

$$FeSO_4.7H_2O \qquad\qquad Na_2CO_3.10H_2O$$
Crystallized Ferrous Sulphate or Green Vitriol.   Crystallized Sodic Carbonate or Sal Soda.

The same salt, when crystallized, at different temperatures not unfrequently combines with different amounts of water of crystallization, the less amounts corresponding to the higher temperatures. Thus manganous sulphate may be crystallized with three different amounts of water of crystallization. We have

$MnSO_4.7H_2O$ when crystallized below 6° C.
$MnSO_4.5H_2O$ " " between 7° and 20°.
$MnSO_4.4H_2O$ " " between 20° and 30°.

The crystalline forms of these three compounds are entirely different from each other; and this fact again corroborates the view that the molecules of water, while a part of the crystalline structure, are not a part of the chemical type of the salt. It will be well to distinguish the molecular aggregate, which the symbols of this section represent, from the simpler chemical molecules by a special term, and we propose to call them crystalline molecules. While, however, there is little room for difference of opinion in regard to the relations in which the molecules of water stand to the structure of most crystals, there are cases where the condition is apparently far less simple, and where we find the water so firmly bound to the salt itself that it seems to form a part of its atomic structure.

Questions and Problems.

1. Analyze reactions [42]. Show what is meant by a metallic hydrate, and define the term alkali. Write the similar reactions which may be obtained with lithium, calcium, and rubidium. Name in each case the class of compounds to which the factors and products belong. Also represent these reactions by graphic symbols.

2. Analyze reactions [43]. State the distinction between an alkaline earth and an alkali, and write the similar reactions which may be obtained with barium and strontium. Name in each case

96   BASES, ACIDS, AND SALTS.

the class of compounds to which the factors and products belong. Also represent the reactions by graphic symbols.

3. Analyze reactions [44] and [45], and write the similar reactions which may be obtained with either of the metals, calcium, strontium, barium, and magnesium. What theory of the constitution of the metallic hydrates do these reactions suggest?

4. In what respects do the hydrates $Ca = O_2 = H_2$ and $Mg = O_2 = H_2$ differ from $K\text{-}O\text{-}H$ and $Na\text{-}O\text{-}H$?

5. Analyze reactions [46], and show that the principal products must be regarded as hydrates. Name the class of compounds to which the other products and factors belong.

6. State the third theory which is held in regard to the constitution of the hydrates, and write the symbols of the different hydrates according to this view. Also bring these symbols into comparison with those of the same compounds written after the other two plans, and show by means of graphic symbols how far these forms are arbitrary, and how far they represent fundamental differences.

7. In what sense may the solution of ammonia gas in water be regarded as an hydrate? Write reactions [46], using ammonic hydrate instead of the hydrates of sodium, potassium, and barium.

8. In what relation do the metallic oxides stand to the hydrates? Define the term *base*.

9. Define the term *salt*, and illustrate your definition by examples.

10. Define the term *acid*. How does an acid differ from a metallic hydrate? Is an acid necessarily an hydrate? What two classes of acids may be distinguished?

11. What is the distinction between an acid and a basic radical. How are they related to the two hydrogen atoms of water? Assuming that there is no difference between these two atoms in the original molecule of water, does not the replacement of one of the atoms by a radical of either class alter the relations of the second? Is there not an analogy between these phenomena and those of magnetism?

12. Analyze reactions [47 et seq.], and point out the evidence of acidity in each case.

13. Analyze the following reactions.

$$K\text{-}O\text{-}H + HF = KF + H_2O$$

$$Ca = O_2 = H_2 + H_2 = O_2 = CO = Ca = O_2 = CO + 2H_2O$$

$$Cu = O_2 = H_2 + 2H\text{-}O\text{-}NO_2 = Cu = O_2 = (NO_2)_2 + 2H_2O$$

BASES, ACIDS, AND SALTS. 97

$$NaCl + H_2=O_2=SO_2 = H, Na=O_2=SO_2 + HCl$$
$$2NaCl + H_2=O=SO_2 = Na_2=O_2=SO_2 + 2HCl.$$

Point out the different acids and bases. In what does the evidence of their acidity or basicity appear either in these or in reactions previously given? Show in each case how the replacement of the hydrogen atoms is obtained, and illustrate the difference between the hydrogen atoms of an acid and those of a base. What two classes of acids may be distinguished?

14. Regarding the hydrates as compounds of hydroxyl, how can you define the acids and bases of this class?

15. Represent the composition of nitric, sulphuric, and phosphoric acid by graphic symbols, and show that the two modes of writing their symbols embody essentially the same idea.

16. Hydrochloric acid, acetic acid, nitric acid, hydriodic acid, hydrobromic acid, sulphuric acid, carbonic acid, and phosphoric acid have what basicity? Point out, in the various reactions given in this chapter, the evidence in each case, and write the symbols of the possible sodic salts of the different acids.

17. What class of compounds do the symbols SO_3, N_2O_5, P_2O_5, CO_2, and SiO_2 represent? By a comparison of symbols show how these compounds may be regarded as formed from water, and how they are related to the corresponding acids. To what class of compounds do they stand in direct antithesis?

18. Define the terms basic and acid hydrate; basic and acid anhydride, and compare reactions [49] with [44 and 45].

19. Analyze the reaction, $BaO + SO_3 = BaO, SO_3$.

What reason may be urged for writing the symbol of baric sulphate in this way? What was the theory of the dualistic system in regard to such compounds? Represent the symbol by the graphic method, and seek to determine whether the dualistic form is compatible with the theory of molecular unity.

20. The following symbols represent compounds of what class?

$H\text{-}O\text{-}H$; $H_3\equiv O_3\equiv PO$; $Fe=O_2=H_2$; $2H\text{-}(HO)$; $(PO_2)_2=O$;

$K\text{-}O\text{-}H$; $Ca=O_2=H_2$; $C_2H_5\text{-}O\text{-}H$; $2Na\text{-}O\text{-}H$; $(C_5H_9O)_2=O$;

$H_4\equiv O_4\equiv Si$; $H\text{-}O\text{-}NO_2$; $H_2=O_2=SO_2$; $(Fe\text{-}Fe)\equiv O_3$; $H\text{-}O\text{-}C_2H_3O$;

$Ca_2\equiv O_4\equiv Si$; $K\text{-}O\text{-}NO_2$; $(C_4H_9)_2=O$; $Na_2=O_2=SO_2$; $C_2H_5\text{-}O\text{-}C_2H_3O$.

98   BASES, ACIDS, AND SALTS.

Give in each case the name of the compound so far as you are able to infer it from examples previously given, and show how the symbol is related to that of water.

21. Point out the acid basic and neutral salts among the compounds represented by the following symbols: —

$H, Na = O_2 = CO$ $H, K = O_2^=(C_2O_2)$ $(Hg\text{-}O\text{-}Hg\text{-}O\text{-}Hg) = O_2^=SO_2$

$Na_2^=O_2 = CO$ $K_2^=O_2^=(C_2O_2)$ $[Hg\text{-}Hg]^=O_2^=(NO_2)_2$

$H_2, Cu \equiv O_4 \equiv Si$ $Cu = O_2^=(NO_2), H$ $[Fe\text{-}Fe]^{\equiv}_{\equiv}O_3^{\equiv}(SO_2)_3$

$Bi \equiv O_3^=(NO_2), H_2$ $H_2, K \equiv O_3 \equiv As$ $K_2^=O_2^=(SO_2\text{-}O\text{-}SO_2)$.

What two classes of basic salts may be distinguished? Convert the symbols into the dualistic form.

22. Analyze reactions [49 and 50], and show how far they justify the dualistic form given to the symbols. Represent the same reactions in the typical form.

23. What class of compounds do the following symbols represent?

$Ag_3 = S_3^= As$ $Ag\text{-}S\text{-}SbS$ $Ca = S_2^= H_2$.

Write the symbols of the corresponding oxygen compounds.

24. Explain the theory of the colored test papers, and the use of the terms acid and basic in connection with them. To what confusion does the double meaning of these terms sometimes lead?

25. The members of the series of alcohols stand in what relation to each other? Does the same relation exist between the members of the series of fat acids, glycols, &c.? Find a general symbol, which will represent the composition of each of these classes of compounds.

26. In what relations do the alcohols stand to the fat acids, and the glycols to the acids derived from them?

27. Select examples from each of the classes of compounds described in sections 40, 41, and 42, and bring the symbols into comparison with those of some simple hydrate or anhydride with which they exactly correspond.

28. We are acquainted with a class of compounds known as condensed glycols, one of which has the following symbols: —

$(C_2H_4\text{-}O\text{-}C_2H_4\text{-}O\text{-}C_2H_4) = O_2^= H_2$.

BASES, ACIDS, AND SALTS.

To what class of salts does this correspond?

29. Judging from the following symbols of a few of the salts of tartaric acid, what conclusion should you reach in regard to the atomicity and basicity of this acid?

$$H_4\equiv O_4\equiv(C_4H_2O_2)\ ;\quad K, H_3\equiv O_4\equiv(C_4H_2O_2)\ ;\quad K_2, H_2\equiv O_4\equiv(C_4H_2O_2)\ ;$$

$$(C_2H_5)_2, H_2\equiv O_4\equiv(C_4H_2O_2)\ ;\quad (C_2H_5)_2, (C_2H_3O)_2\equiv O_4\equiv(C_4H_2O_2)$$

30. What is the atomicity and basicity of the different acids whose symbols have been given in this chapter? Does the basicity of the different hydrocarbon acids (§ 40 to § 43) appear to have any connection with the number of oxygen atoms in the radical?

31. How do you explain the state of combination of the water which enters into the composition of most crystalline salts? Show by an example how this mode of combination is represented symbolically. What facts may be adduced in support of the opinion that the molecules of water are not a part of the chemical type of the salt.

CHAPTER X.[1]

CHEMICAL NOMENCLATURE.

45. *Origin of Nomenclature.* — Previous to the year 1787 the names given to chemical compounds were not conformed to any general rules; and many of these old names, such as *oil of vitriol, calomel, corrosive sublimate, red precipitate, saltpetre, sal-soda, borax, cream of tartar, Glauber's* and *Epsom salts,* are still retained in common use. As chemical science advanced, and the number of known substances increased, it became important to adopt a scientific nomenclature, and the system which came into use was due almost entirely to Lavoisier, who reported to the French Academy on the subject, in behalf of a committee, in the year named above. In the Lavoisierian nomenclature the name of a substance was made to indicate its composition; and at the time of its adoption, and for fifty years after, it was probably the most perfect nomenclature which any science ever enjoyed. It was based, however, on the dualistic theory, of which Lavoisier was the father; and, when at last the science outgrew this theory, the old names lost much of their significance and appropriateness. Within the last few years the English chemists have attempted to modify the old nomenclature so as to better adapt the names to our modern ideas. Unfortunately the result, like most attempts to mend a worn-out garment, is far from satisfactory, although it is probably the best which under the circumstances could be attained. The new nomenclature has not the simplicity or unity of the old, and its rules cannot be made intelligible until the student is more or less acquainted with the modern chemical theories. Fortunately, however, the admirable system of chemical symbols supplies the defects of the nomenclature, and for many

---

[1] In studying this chapter, the student is expected to remember the names corresponding to the different symbols, and also the symbols corresponding to the names.

purposes may be used in its place. We have, therefore, developed this system first, but have also used, meanwhile, the corresponding scientific names, so that the student might become familiar with the nomenclature, and gather its rules as he advanced. A brief summary of these rules is all that will be necessary here.

46. *Names of Elements.* — The names of the elements are not conformed to any fixed rules. Those which were known before 1787, such as sulphur, phosphorus, arsenic, antimony, iron, gold, and the other useful metals, retain their old names. Several of the more recently discovered elements have been named in allusion to some prominent property or some circumstance connected with their history: as *oxygen*, from ὀξὺς γεννάω (acid-generator); *hydrogen*, from ὕδωρ γεννάω (water-generator); *chlorine*, from χλωρός (green); *iodine*, from ἰωδής (violet); *bromine*, from βρῶμος (fetid odor). The names of the newly discovered metals have a common termination, *um*, as *potassium, sodium, platinum;* and the names of several of the newly discovered metalloids end in *ine*, as *chlorine, bromine, iodine, fluorine*. Equally arbitrary names have been given to the compound radicals; but, with a few exceptions, they all terminate in *yl* or *ene*, as *ethyl, acetyl, hydroxyl*, and *ethylene, acetylene*, &c.

47. *Names of Binary Compounds.*[1] The simple compounds of the elements with oxygen are called oxides, and the specific names of the different oxides are formed by placing before the word "oxide" the name of the element, but changing the termination into *ic* or *ous*, to indicate different degrees of oxidation, and using the Latin name of the element in preference to the English, both for the sake of euphony and in order to secure more general agreement among different languages. When the same element unites with oxygen in more than two proportions, the Latin prepositions or numeral adverbs, *sub, per, bis*, &c., are prefixed to the word "oxide," in order to indicate the additional degrees. Formerly these compounds were called oxides of the different elements, the degrees of oxidation being indicated solely by the prefixes; and, as the old names are still in very general use, they are also given in the following examples: —

[1] Compounds of two elements.

102   CHEMICAL NOMENCLATURE.

	New Names.		Old Names.
AgO	is Argentic Oxide	or	Oxide of Silver
N_2O	" Nitrous Oxide	"	Protoxide of Nitrogen
NO	" Nitric Oxide	"	Deutoxide of Nitrogen
NO_2	" Nitric Peroxide	"	Peroxide of Nitrogen
FeO	" Ferrous Oxide	"	Protoxide of Iron
Fe_2O_3	" Ferric Oxide	"	Sesquioxide of Iron.

An important exception to the above rules is made in the case of those oxides which, when combined with the elements of water, form acids. As has been already stated, page 61, such compounds are called anhydrides, but the degrees of oxidation are distinguished as before, thus:—

	New Names.		Old Names.
SO_2	is Sulphurous Anhydride	or	Sulphurous Acid
SO_3	" Sulphuric Anhydride	"	Sulphuric Acid
N_2O_3	" Nitrous Anhydride	"	Nitrous Acid
N_2O_5	" Nitric Anhydride	"	Nitric Acid
P_2O_3	" Phosphorous Anhydride	"	Phosphorous Acid
P_2O_5	" Phosphoric Anhydride	"	Phosphoric Acid
CO_2	" Carbonic Anhydride	"	Carbonic Acid
SiO_2	" Silicic Anhydride	"	Silicic Acid.

The names in common use, even among chemists, of the earths, the alkaline earths, and the alkaline oxides, make another important exception to the general rules given above, thus:—

Al_2O_3	Aluminic Oxide is commonly called	Alumina
BaO	Baric Oxide " " "	Baryta
SrO	Strontic Oxide " " "	Strontia
CaO	Calcic Oxide " " "	Lime
MgO	Magnesic Oxide " " "	Magnesia
K_2O	Potassic Oxide " " "	Potassa
Na_2O	Sodic Oxide " " "	Soda.

As this last class of oxides stands in the same relation to the bases in which the previous class stands to the acids, they have also been called by some chemists anhydrides.

The names of the binary compounds of the other elements are formed like those of the oxides.

CHEMICAL NOMENCLATURE. 103

Compounds of Chlorine are called Chlor*ides*
" " Bromine " " Brom*ides*
" " Iodine " " Iod*ides*
" " Fluorine " " Fluor*ides*
" " Sulphur " " Sulph*ides*
" " Nitrogen " " Nitr*ides*
" " Phosphorus " " Phosph*ides*
" " Arsenic " " Arsen*ides*
" " Antimony " " Antimon*ides*
" " Carbon " " Carbon*ides*.

Moreover, the specific names of the several compounds also follow the analogy of the oxides, thus:—

New Names.   *Old Names.*

$SnCl_2$ is Stannous Chloride or Protochloride of Tin
$SnCl_4$ " Stannic Chloride " Perchloride of Tin
Fe_2S " Diferrous Sulphide " Subsulphide of Iron
FeS " Ferrous Sulphide " Protosulphide of Iron
Fe_2S_3 " Ferric Sulphide " Sesquisulphide of Iron
FeS_2 " Ferric Disulphide " Bisulphide of Iron
$CaFl_2$ " Calcic Fluoride " Fluoride of Calcium.

Here, again, must be noticed several exceptions to the general rule. Several simple compounds of the elements with hydrogen, of which the hydrogen is easily replaced with a metal or positive radical, are called acids, and retain the specific names of the old nomenclature, thus:—

HCl or Hydric Chloride is called Hydrochloric Acid
HBr " Hydric Bromide " " Hydrobromic Acid
HI " Hydric Iodide " " Hydriodic Acid
HFl " Hydric Fluoride " " Hydrofluoric Acid
H_2S " Hydric Sulphide " " Hydrosulphuric Acid.

The last compound is frequently called also sulphuretted hydrogen, and several other hydrogen compounds are named after the same analogy, while others again are always called by their well-known trivial names, thus:—

H_3Sb is Antimoniuretted Hydrogen
H_3As " Arseniuretted Hydrogen
H_3P " Phosphuretted Hydrogen
H_3N " Ammonia Gas
H_4C " Marsh Gas or Light Carburetted Hydrogen
H_4C_2 " Olefiant Gas or, as a radical, Ethylene.

48. *Ternary Compounds.* — Of the old class of ternary compounds, it is only those which are formed after the type of water for which the rules of the nomenclature need at present be explained.

49. *Bases.* — These we call simply hydrates, and for the specific name we take the name of the positive radical, changing the termination into *ic* or *ous*, and using such prefixes as circumstances may require, thus : —

	New Names.	Old Names.
$K\text{-}O\text{-}H$	is Potassic Hydrate or	Hydrate of Potassa
$Ca=O_2=H_2$	" Calcic Hydrate "	Hydrate of Lime
$Fe=O_2=H_2$	" Ferrous Hydrate "	{Hydrate of Protoxide of Iron.
$Fe_2^{\equiv}O_3^{\equiv}H_6$	" Ferric Hydrate "	{Hydrate of Sesquioxide of Iron.
$(O=Fe\text{-}Fe=\overset{\text{II}}{O})=O_2=H_2$	" Diferric Hydrate, the mineral Göthite.	

50. *Acids.* — The inorganic acids all take their specific names from the name of the most characteristic element of the negative radical, which is modified by terminations and prefixes as before, only the last are usually taken from the Greek rather than the Latin. Here the old and the new names coincide.

$H\text{-}O\text{-}NO_2$	is called	Nitric Acid
$H_2=O_2=SO_2$	" "	Sulphuric Acid
$H_2=O_2=SO$	" "	Sulphurous Acid
$H_2=O_2=(S\text{-}O\text{-}\overset{\text{II}}{S})$	" "	Hyposulphurous Acid

The specific names of the organic acids are, as a rule, arbitrary, like tartaric acid, citric acid, malic acid, gallic acid, uric acid, and the like.

51. *Salts.* — The name of a salt is formed from the name of the acid from which the salt is derived, preceded by the names of the basic radicals. When the name of the acid ends in *ic* the termination is changed into *ate*, when in *ous* into *ite*. Moreover, the terminations *ous* and *ic* are retained in connection with the name of the basic radical, and such prefixes are used as may be necessary for distinction, thus : —

CHEMICAL NOMENCLATURE. 105

	New Names.		Old Names.
$Ca = O_2 = CO$	is Calcic Carbonate	or	Carbonate of Lime
$Ca = O_2 = (S\text{-}O\text{-}\overset{\text{II}}{S})$	" Calcic Hyposulphite	"	Hyposulphite of Lime
$Ba = O_2 = SO$	" Baric Sulphite	"	Sulphite of Baryta
$Fe = O_2 = SO_2$	" Ferrous Sulphate	"	Protosulphate of Iron
$Fe_2\equiv O_3\equiv(SO_2)_3$	" Ferric Sulphate	"	Persulphate of Iron
$(NH_4), Mg \equiv O_3 \equiv PO$	" Ammonio-magnesic Phosphate		
$H, (NH_4), Na \equiv O_3 \equiv PO$	" Hydro-ammonio-sodic Phosphate.		

The terms "acid" and "basic" have been used *as parts of the name* of a salt very confusedly. We would propose to limit this special use of these words to such salts as still contain atoms of hydrogen, replaceable by a radical, basic in the first case and acid in the other. This use has been followed on page 87, where the distinction has been pointed out between salts of this class and those basic and acid salts which may be regarded as formed by the cementing together of several radicals into a single complex group. Salts of this last kind we would distinguish by appropriate prefixes, but as examples of names of both forms have already been given on the page cited, it will be unnecessary to multiply them here.

Questions and Problems.

1. Give the names of the compounds represented by the following symbols: —

a. KCl; K_2O; K_2S; $K_2 = O_2 = SO$; $K_2 = O_2 = SO_2$; $K_2 = O_2 = (S\text{-}O\text{-}S)$;

b. FeO; $Fe = O = H$; $Fe = O = CO$; $Fe = O_2 = C_2O_2$; $[Fe_2]\equiv O_3$; $Fe_2\equiv O_6\equiv H_6$; $[Fe_2]\equiv O_6\equiv(NO_2)_6$.

c. $H\text{-}Cl$; $H\text{-}F$; $H\text{-}O\text{-}NO_2$; $H\text{-}O\text{-}NO$; $H_2 = O_2 = SO_2$; $H_2 = O_2 = SO$; $H_3 = O_3 = PO$.

d. $Hg = Cl_2$; $[Hg_2] = Cl_2$; $Cu = S$; $[Cu_2] = S$; PbI_2; KBr; $[Al_2]\equiv O_3$; ZnO.

e. $H,K\text{=}O_2\text{=}SO_2$; $H_2,Na_2\text{≡}O_3\text{≡}PO$; $H,Na\text{=}O_2\text{=}CO$; $H,K\text{=}O_2\text{=}(C_2O_2)$.

f. $N\text{-}N$; N_2O; NO; NO_2; N_2O_3; N_2O_5; MnO; Mn_2O_3; Mn_3O_4; MnO_2.

2. Write the symbols of the following compounds:

 a. Calcic Sulphide; Calcic Sulphite; Calcic Hyposulphite; Calcic Sulphate; Calcic Hydrate; Calcic Sulphohydrate; Calcic Carbonate; Calcic Sulphocarbonate; Calcic Silicate.

 b. Water; Potassic Hydrate; Nitric Acid; Potassic Nitrate; Nitric Anhydride; Potassic Oxide.

 c. Magnesic Oxide; Magnesic Hydrate; Magnesic Nitrate; Magnesic Carbonate; Magnesic Phosphate; Ammonio-magnesic Phosphate.

 d. Cuprous Chloride; Cupric Chloride; Ferrous Chloride; Ferric Chloride; Sulphurous Anhydride; Sulphuric Anhydride; Phosphorous Anhydride; Phosphoric Anhydride.

 N. B. Examples like the above should be greatly multiplied by the teacher, pains being taken to group together the names and symbols in the way best calculated to exhibit their relations and to assist the memory.

CHAPTER XI.

SOLUTION AND DIFFUSION.

52. *Solution.* — The solvent power of water is one of the most familiar facts of common experience, and all liquids possess the same power to a greater or less degree, but they differ very widely from each other in the manifestation of their solvent power, which for each liquid is usually limited to a certain class of solids. Thus mercury is the appropriate solvent of metals, alcohol of resins, ether of fats, and water of salts and of similar compounds of its own type. Water is by far the most universal solvent known, and for this reason, as well as on account of its very wide diffusion in nature, it becomes the medium of most chemical changes. The phenomena of aqueous solution form therefore a very important subject of chemical inquiry, and these alone will be considered in this connection.

The solvent power of water, even on bodies of its own type, differs very greatly. Some solids, like potassic carbonate, or calcic chloride, liquefy in the atmosphere by absorbing the moisture it contains. Such salts are said to *deliquesce*, and are rendered liquid by a very small proportion of water. Other salts, like calcic sulphate, require for solution several hundred times their weight of water, and others again, like baric sulphate, are practically insoluble.

As a general rule the solvent power of water increases with the temperature; but here, again, we observe the greatest differences between different substances. While the solubility of some salts increases very rapidly with the temperature, that of others increases not at all, or only very slightly; and there are a few which are actually more soluble in cold water than in hot. The solubility of each substance is absolutely definite for a given temperature, and we can determine by experiment the exact amount which 100 parts of water will in any case dissolve. The results of such experiments are best represented to the eye by means of a curve drawn as in the accompanying figure on the principles of analytical geometry.

108 SOLUTION AND DIFFUSION.

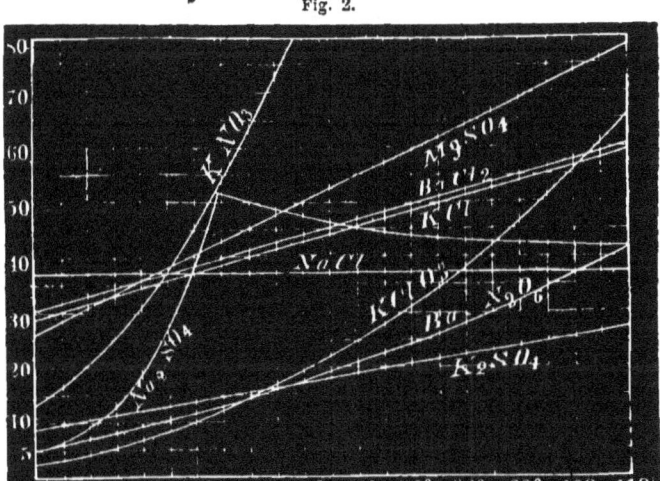

Fig. 2.

The figures on the horizontal line indicate degrees of temperature, and those on the vertical line parts of salt soluble in 100 parts of water. To find the solubility of any salt, for a stated temperature, the curve being given, we have only to follow up the vertical line corresponding to the temperature until it reaches the curve, and then, at the end of the horizontal line which intersects the curve at the same point, we find the number of parts required. These curves also show in each case the law which the change of solubility obeys.

When a liquid has dissolved all of a solid that it is capable of holding at the temperature, it is said to be saturated; but when saturated with one solid the liquid will still exert a solvent power over others; indeed, in some cases the solvent power is thereby increased. When several salts are dissolved together in water, a definite amount of metathesis seems always to take place, and the different positive radicals are divided between the several acids in proportions which depend on the relative strength of their affinities, and on the quantities of each present. If in this way either an insoluble or a volatile product is formed, the solid or the gas at once falls out of the solution, and, the equilibrium being thus destroyed, a new metathesis takes place, and this goes on so long as any of these products can be formed. Here, then, we find a simple explanation of the two important laws already stated on page 37.

53. *Solution of Gases.*— Most liquids, but especially water and alcohol, exert on gases a greater or less solvent power, which is marked by differences of manifestation similar to those we have already studied in the case of solids, although the peculiar physical conditions of the gas somewhat modify the result. Under the same conditions, the volume of gas dissolved is always the same; but it varies with the pressure of the gas on the surface of the liquid, with the temperature, and with the peculiar nature of the gas and the absorbing liquid. The quantity[1] of gas dissolved by a liquid on which it exerts a pressure of 76 c. m. is called the coefficient of absorption. This coefficient, in almost every instance, diminishes with the temperature; but, as in the case of solids, each substance obeys a law of its own, which must be determined by experiment. The observed values at different temperatures for several of the best known gases, when absorbed by water and alcohol, are given in the Chemical Physics, Table VII. With these data we can easily calculate the quantity of any of these gases which a given volume of water or alcohol will absorb, assuming that the gas exerts on the liquid a pressure of 76 c. m. Moreover, since *the quantity of a gas absorbed by a liquid varies directly as the pressure which the gas exerts upon it*, we can easily calculate from the first result the quantity absorbed at any given pressure. Again, it is a direct consequence of the last principle that at a fixed temperature a given mass of liquid will dissolve the *same volume* of gas, whatever may be the pressure. Lastly, if a mass of liquid is exposed to an atmosphere of mixed gases, it will absorb of each the same quantity as if this gas was alone present and exerting on the liquid the same partial pressure which falls to its share in the atmosphere. The amount dissolved of each gas is easily calculated when the partial pressure and the coefficient of absorption are known. It is thus that water absorbs the oxygen and nitrogen gases of our terrestrial atmosphere; and the fact that these two gases are found dissolved in the ocean in very different proportions from those present in the atmosphere is a conclusive proof that the air is a mixture, and not, as was formerly supposed, a chem-

54. Solution and Chemical Change.— There seems at first sight to be a wide difference between solution and chemical change; for, while in the first the solid body becomes diffused through the liquid menstruum without losing its chemical identity or destroying that of the liquid, there is in the second a complete identification of the combining substances in the resulting compound.

The same wide difference appears also between mechanical and chemical solution, which are sometimes confounded by students, because, unfortunately, the same term has been applied to both. When salt or sugar is dissolved in water, the differences between salt and solvent are preserved; but when chalk is dissolved in hydrochloric acid, or copper in nitric acid, there is a complete *identification of the differences* in the resulting compound; and the only ground for calling such chemical changes solution is the fact that the solution of the resulting salt in the water, used as the medium of the chemical change, is frequently an essential condition of the process.

But if, instead of comparing extreme cases, we study the whole range of chemical phenomena, we shall find that the distinction is by no means so clearly marked. In many cases what seems to be a simple solution can be shown to be a mixed effect at least of solution and chemical combination; and between this condition of things, where the evidence of chemical combination is unmistakable, and a simple solution, like that of sugar in water, we have every degree of gradation. To such an extent is this true, that the facts seem to justify the opinion that solution is in every case a chemical combination of the substances dissolved with the solvent, and that it differs from other examples of chemical change only in the weakness of the combining force.

The metallic alloys afford another striking illustration of the same principle. They are originally solutions of one metal in another; but in many cases the result is greatly modified by the chemical affinities of the metals and their tendency to form definite chemical compounds.

55. *Liquid Diffusion*.— Closely connected with the phenomena of solution are those of liquid diffusion. These phenomena may be studied in their simplest form, by placing an open vial filled with a solution of some salt in a much larger jar of pure water, as shown in Fig. 3, and so carefully arranging the details of the experiment that the surfaces of the two liquids may be brought in contact without mixing them mechanically. It will then be found that the salt molecules will slowly escape from the vial and spread throughout the whole volume of the water. The rate of the diffusion increases with the temperature equally for all substances, and the whole phenomenon is probably caused by that same molecular motion to which we refer the effects of heat. At best, however, the diffusion is very slow, as we should expect, considering the limited freedom of motion which the liquid molecules possess. It is found, also, that the rate of diffusion differs very greatly for the different soluble salts; but these may be divided into groups of equidiffusive substances, and the rates of diffusion of the several groups bear to each other simple numerical ratios. If a mixture of salts be placed in the vial, it is found that the presence of one salt affects to some degree the diffusion of the other; but if the difference of rate is considerable, a partial separation may be effected, and even weak chemical compounds may be thus decomposed.

Fig. 3.

56. *Crystalloids and Colloids.* — There is a very great difference of diffusive power between the ordinary crystalline salts (including most of the common acids and bases) and such substances as gum, caramel, gelatine, and albumen, which are incapable of crystallizing, and which give insipid viscid solutions, readily forming into jelly; hence the name colloids, from κόλλη, glue. The last class is distinguished by a remarkable sluggishness and indisposition to diffusion; as is illustrated by the fact that sugar, one of the least diffusible of the crystalloids, diffuses seven times more rapidly than albumen, and fourteen times more rapidly than caramel. Our theories would lead us to believe that this great difference of diffusive power is caused by the fact that the molecules of colloids are far more complex atomic aggregates than those of crystalloids, and therefore are heavier and move more slowly. Moreover, the diffusive power

is only one of many characters which point to a great molecular difference between these two classes of substances.

57. *Dialysis*. — The difference of diffusive power between the two classes of compounds distinguished in the last section is still further increased when the aqueous solution is separated from the pure water by some colloidal membrane, and upon this fact Professor Graham of London, to whom we owe our whole knowledge of this subject, has based a simple method of separating crystalloids from colloids, which he calls dialysis.

A shallow tray is prepared by stretching parchment paper (which is itself an insoluble colloid) over one side of a guttapercha hoop, and holding it in place by a somewhat larger hoop of the same material. The solution to be *dialysed* is poured into this tray, which is then floated on pure water whose volume should be eight or ten times greater than that of the solution. Under these conditions the crystalloids will diffuse through the porous septum into the water, leaving the colloids on the tray, and in the course of two or three days a more or less complete separation of these two classes of substances will have taken place.

In this way arsenious acids and similar crystalloids may be separated from the colloidal materials, with which, in cases of poisoning, they are frequently found mixed in the stomach; and by an application of the same method alumina, ferric oxide, chromic oxide, stannic, metastannic, titanic, molybdic, tungstic, and silicic acids have all been obtained dissolved in water in a colloidal condition. All these substances usually exist in a crystalline condition. The colloidal condition appears to be an abnormal state, and in almost all such substances there is a tendency towards the crystalloid form.

58. *Diffusion of Gases*. — Gases diffuse much more rapidly than liquids, as we should naturally expect from the greater freedom of motion which their molecules possess. Moreover, if the theory of the molecular condition of gases is correct, we ought to be able to calculate the relative rates of diffusion of different gases from their respective molecular weights. If it is true, as stated on page 11, that at any given temperature

$$\tfrac{1}{2} m V^2 = \tfrac{1}{2} m' V'^2$$

then it follows that

$$V : V' = \sqrt{\tfrac{1}{2} m'} : \sqrt{\tfrac{1}{2} m} = \sqrt{\text{Sp. Gr}'.} : \sqrt{\text{Sp. Gr.}}$$

Hence, if two masses of gas are in contact, the molecules of either gas must move into the space filled by the other with velocities which are inversely proportional to the square roots of the respective specific gravities. If one gas is hydrogen (Sp. Gr. $= 1$), and the other oxygen (Sp. Gr. $= 16$), the molecules of hydrogen must move past the section separating the two masses four times as rapidly as those of oxygen; and, since all gas molecules occupy the same volume, it follows further that four volumes of hydrogen must enter the space filled by the oxygen, while one volume of oxygen is passing in the opposite direction Numerous experiments have fully confirmed this theoretical deduction, and the close agreement between theory and experiment furnishes important evidence in favor of the theory itself. Such experiments can be made, moreover, with great accuracy, since the molecular motion is not arrested by various porous septa, which may be used to separate the two masses of gas, and which entirely prevent the passage of gas currents that might otherwise vitiate the results.

CHAPTER XII.

COMBUSTION.

59. *The Atmosphere.* — The earth is surrounded by an ocean of aeriform matter called the atmosphere, and many of the most important chemical changes which we witness in nature are caused by the reaction of this atmosphere on the substances which it surrounds and bathes. The great mass of the atmosphere consists of the two elementary gases, oxygen and nitrogen, mixed together in the proportions indicated in the following table : —

Air contains.	Composition By Volume.	Composition By Weight.
Oxygen,	20.96	23.185
Nitrogen,	79.04	76.815
	100.	100.

That the air is a mixture, and not a chemical compound, is proved by the action of solvents upon it (§ 50) ; but, nevertheless, the analyses of air collected in different countries, and at different heights in the atmosphere, show a remarkable constancy in its composition. Besides these two gases, which make up over 93 per cent of its whole mass, the air always contains variable quantities of aqueous vapor, carbonic anhydride, and ammonia, and sometimes also traces of various other gases and vapors.

60. *Burning.* — Of the two chief constituents of the atmosphere, nitrogen gas is a very inert substance, and serves chiefly to restrain its more energetic associate. Oxygen gas, on the other hand, is endowed with highly active affinities, and tends to enter into combination with other elementary substances, and with many compounds which are not already saturated with this all-pervading element. Many of these substances, such as phosphorus, sulphur, petroleum, coal, and wood, have such a strong affinity for oxygen, that, under certain conditions, they will absorb it from the atmosphere, and combine with it

under the evolution of heat and light. These substances are said to be combustible, and the process of combination is called combustion. Moreover, all burning with which we are familiar in common life consists in the union of the burning body with the oxygen of the air. The chemical process in these cases may be expressed, like any other chemical reaction, in the form of an equation.

Burning of Hydrogen Gas.

$$2\,\underset{\text{Hydrogen Gas.}}{H\text{-}H} + O\text{=}O = 2\,\underset{\text{Aqueous Vapor.}}{H_2O}.\qquad [53]$$

Burning of Carbon (Charcoal).

$$\underset{\text{Carbon.}}{C} + O\text{=}O = \underset{\text{Carbonic Anhydride.}}{CO_2}.\qquad [54]$$

Burning of Benzole.

$$2\,\underset{\text{Benzole.}}{C_6H_6} + 15\,O\text{=}O = 12\,CO_2 + 6\,H_2O.\qquad [55]$$

Burning of Alcohol.

$$\underset{\text{Alcohol.}}{C_2H_6O} + 3\,O\text{=}O = 2\,CO_2 + 3\,H_2O.\qquad [56]$$

Burning of Sulphur.

$$S\text{-}S + 2\,O\text{=}O = 2\,\underset{\text{Sulphurous Anhydride.}}{SO_2}.\qquad [57]$$

Burning of Phosphorus.

$$P_2\text{=}P_2 + 5\,O\text{=}O = 2\,\underset{\text{Phosphoric Anhydride.}}{P_2O_5}.\qquad [58]$$

Burning of Magnesium.

$$2\,Mg + O\text{=}O = 2\,\underset{\text{Magnesic Oxide.}}{MgO}.\qquad [59]$$

The four substances, hydrogen gas, charcoal, benzole, and alcohol, may be regarded as types of our ordinary combustibles; and, as the first four reactions show, the products of their combustion are aeriform. Moreover, these products are wholly devoid of any sensible qualities, and hence the apparent annihilation of the burning substance, and the reason that for so long a period the nature of the process remained undiscovered. That these qualities of the products of ordinary combustion are not necessary conditions of the process, but remarkable adaptations in the properties of those combustibles which are our artificial sources of light and heat, is shown by the fact, that, in the last two reactions, the products of the combustion are solids, while in [57] the product is a noxious suffocating gas.

A careful inspection of the reactions will also teach the student several other important facts in regard to the processes here represented. It will be seen that, in the burning of hydrogen gas, two volumes of hydrogen gas and one volume of oxygen gas combine to form two volumes of aqueous vapor. It will further be noticed, that, in the burning of carbon and of sulphur, a given volume of oxygen gas yields in each case its own volume of the aeriform product. The carbon in the one case, and the sulphur in the other, are absorbed, as it were, by the gas, without any increase of volume. Further, if the experiments are made, which these reactions represent, it will appear that, in all those cases where the combustible is represented as a gas, the combustion is accompanied by flame, while in the case of carbon, which is a fixed solid, there is no proper flame. Hence we learn that flame is burning gas, and that only those substances burn with flame which are either gases themselves, or which, at a high temperature, become volatilized, or generate combustible vapors. Still other important facts connected with the process of combustion will be learned by solving the following problems according to the rules already given (§§ 24 and 25).

Problem. How many cubic centimetres of hydrogen gas, and how many of oxygen gas, are required to form one cubic centimetre of liquid water?[1] Ans. 1,240 $\overline{cm}^3$ of hydrogen gas, and 620 $\overline{cm}^3$ of oxygen gas.

Problem. How many cubic metres of air are required to burn 448 kilogrammes of coal, assuming that the coal is pure carbon? Ans. 833.333 $\overline{m}^3$ of oxygen gas, or 3,975.83 $\overline{m}^3$ of atmospheric air.

COMBUSTION. 117

Problem. How many cubic metres of carbonic anhydride are formed by the burning of 1,000 kilogrammes of coal, assuming, as before, that the coal is pure carbon? Ans. 1,860.

Problem. How many litres of carbonic anhydride, and how many of aqueous vapor, would be formed by burning one litre of benzole vapor? Ans. Simple inspection of the equation shows that 6 litres of the first and 3 litres of the second would be formed.

Problem. How many litres of carbonic anhydride, and how many of aqueous vapor, would be formed by burning one litre of liquid alcohol (C_2H_6O)? Sp. Gr. of liquid at 0° = 0.815. Ans. One litre of alcohol weighs 815 grammes or 9,097 criths, and, since the Sp. Gr. of alcohol vapor is 23, this quantity of liquid would yield 395.6 litres of vapor. Hence there would be formed $2 \times 395.6 = 791.2$ litres of carbonic anhydride, and $3 \times 395.6 = 1,186.8$ litres of aqueous vapor.

61. *Heat of Combustion.* — The reactions of the last section represent only the chemical changes in the processes of burning. The physical effects which accompany the chemical changes our equations do not indicate, but it is these remarkable manifestations of power which chiefly arrest the student's attention, and on this power the importance of the processes of combustion as sources of heat and light wholly depends.

The immediate cause of the power developed in the process of combustion is to be found in the clashing of material atoms. Urged by that immensely powerful attractive force we call chemical affinity, the molecules of oxygen in the surrounding atmosphere rush, from all directions, and with an incalculable velocity, upon the burning body. The molecules of oxygen thus acquire an enormous moving power; and when, at the moment of chemical union, the onward motion is arrested, this moving power is distributed among the surrounding molecules, and is manifested in the phenomena of heat and light.[1] (Compare § 12.)

[1] According to our best knowledge, the phenomena of light are merely another manifestation of the same molecular motion which causes the phenomena of heat. When we speak of the amount of heat produced, we refer always to the total amount of molecular motion; although, even in the most brilliant illumination, the amount of mechanical power manifested as light appears to be inconsiderable as compared with that which takes the form of heat.

The quantity of heat evolved during combustion varies very greatly with the nature of the combustible employed, but it is always constant for the same combustible if burnt under the same conditions, and is exactly proportional to the weight of combustible consumed. We give in the following table the amount of heat evolved by one kilogramme of several of the most common combustibles when they are burnt in oxygen gas in their ordinary physical state. The numbers represent what is called the calorific power of the combustible. With the exception of the two last, which are only approximate values, they are the results of very accurate experiments made by Favre and Silbermann.

Calorific Power of Combustibles.

	Units.		Units.
Hydrogen,	34,462	Sulphur,	2,221
Marsh Gas,	13,063	Wood Charcoal,	8,080
Olefiant Gas,	11,858	Carbonic Oxide,	2,400
Ether,	9,027	Dry Wood (about),	3,654
Alcohol,	7,184	Bituminous Coal, "	7,500

The calorific power of our ordinary hydrocarbon fuels may be calculated approximately when their composition is known. Most of these combustibles contain more or less oxygen, and it is found, as might be expected, that the amount of heat developed by the perfect combustion of the fuel is equal to that which would be produced by the perfect combustion of all the carbon, and of so much of the hydrogen as is in excess of that required to form water with the oxygen present. The rest of the hydrogen may be regarded, so far as relates to the present problem, as in combination with oxygen in the state of water; and in estimating the available heat produced, we must deduct the amount of heat required to convert, not only this water into steam, but also any hygroscopic water which may be present. Moreover, if we use in our calculation the value of the calorific power of hydrogen given in the table above, we must also deduct the amount of heat required to convert into vapor all the water formed in the process of burning, because, in the experiments by which this value was obtained, the aqueous vapor formed was subsequently condensed to water and gave out its latent heat.

Problem. Given the average composition of air-dried wood as in the table, to find the calorific power.

Carbon,	400	From the results of analysis we easily deduce
Hydrogen,	48	Quantity of H in combination with O 41
Oxygen,	328	" " available as fuel 7
Nitrogen and Ash,	24	Quantity of water formed by burning 48 parts hydrogen } 432
Hygroscopic Water,	200	
	1000	Hygroscopic Water 200
		Total quantity of water evaporated 632

Units of Heat.

400 grammes of carbon yield 3,232
7 " " hydrogen " 241
─────
3,473
Deduct amount of heat required to convert 632 grammes of
    water into vapor. (See § 14.) 339
Calorific power of air-dried wood 3,134

From the mechanical equivalent of heat given on page 11, and from the data of the above table, we can easily calculate the mechanical power developed in ordinary combustion, and the student will be surprised to find how great this power is. The burning of one kilogramme of charcoal produces an amount of heat which is equivalent to $8,080 \times 423 = 3,417,840$ kilogramme metres; that is, the moving power which is developed by the clashing of the atoms during the combustion of this small amount of coal is equal to that which would be produced by the fall of a mass of rock weighing 8,080 kilogrammes over a precipice 423 metres high, and, could this power be all utilized, it would be adequate to raise the same weight to the same height, or to do any other equivalent amount of work. The steam-engine is a machine for applying this very power to produce mechanical results; but, unfortunately, in the best engines we do not utilize much more than $\frac{1}{20}$ of the power of the fuel; and to find a more economical means of converting heat into mechanical effect is one of the great problems of the present age.

62. *Calorific Intensity.* — The calorific intensity of fuel is to be carefully distinguished from its calorific power. By *calorific power* is meant, as we have seen, the total quantity of heat developed by the combustion of a given amount of fuel. By *calorific intensity*, we mean the maximum temperature developed in the process of combustion. Provided the products are the same, the total amount of heat produced in any case is

not materially influenced by the rapidity of the process; but it is evident that the temperature of the burning fuel will depend, other things being equal, on the rapidity with which the heat is developed as compared with the rapidity with which it is dissipated through surrounding objects; and, when the combination with oxygen is very slow, the heat may be dissipated as fast as it is generated, and then the temperature of the burning body will not rise above that of the surrounding atmosphere, as is the case in many of the processes of slow combustion.

Assuming, however, that all the heat is retained by the products of combustion, we can calculate the maximum temperature which can in any case be produced, provided the calorific power of the fuel and the specific heat of the products of combustion are known. The calorific intensity is simply the temperature to which the heat generated by the burning of each portion of the fuel can raise the products of its own combustion. Assume that the quantity burnt is one kilogramme, that the calorific power or number of units of heat produced is C, that the weights of the various products of combustion are W, W', W'', &c., and that the specific heats of these products are S, S', S'', &c. Then $WS + W'S' + W''S'' +$ &c., represents the amount of heat required to raise the temperature of the whole mass of the products one centigrade degree (§ 16), — and the maximum temperature, to which these products can be raised in the process of combustion, must be

$$T = \frac{C}{WS + W'S' + W''S''} \qquad [60]$$

Problem. Find the calorific intensity of charcoal burnt in pure oxygen, and also in air under constant atmospheric pressure.

Solution. By [54] we easily find that each kilogramme of carbon yields, by burning, 3.67 kilogrammes of carbonic anhydride, which is the sole product of its combustion when burnt in pure oxygen. The specific heat of carbonic anhydride (Chem. Phys. 235) is 0.2164. The calorific power of charcoal is 8,080. By substituting these values in [60] we get $T =$ 10,174°.

When the charcoal burns in air, the 3.67 kilogrammes of carbonic anhydride formed by the combustion are mixed with a large mass of inert nitrogen, which must be regarded as one of the products of the combustion. The weight of this nitrogen is easily calculated from the known composition of air by weight (§ 56) and from the amount of oxygen consumed in the process.

23.2 : 76.8 = 2.67 : x; or $x = 2.67 \times 3.31 = 8.84$.

We have now, besides the values given above, $W' = 8.84$ and S_i' the specific heat of nitrogen, equal to 0.244. Whence $T' = 2,738°$.

Problem. Find the calorific intensity of hydrogen gas burnt in oxygen and burnt in air.

Solution. One kilogramme of hydrogen yields 9 kilogrammes of aqueous vapor. The specific heat of aqueous vapor is 0.4805. The calorific power of hydrogen is not so great when the gas is burnt under ordinary conditions as that given in the table on page 118; for in the experiments of Favre and Silbermann the vapor formed by the combustion was subsequently condensed to water, and gave out its latent heat, while in a burning flame of hydrogen no such condensation takes place. Hence $C = 34,462 - (537 \times 9) = 29,629$. We also have $W = 9$ and $S = O, 475$. Whence $T = 6,853°$.

When hydrogen is burnt in air, the nitrogen, mixed with the aqueous vapor, weighs 26.49 kilogrammes and S is the same as in the previous problem. Whence $T' = 2,746°$.

It appears then from these problems, that, although the calorific power of hydrogen is much greater than that of carbon, its calorific intensity is less. But it must be remembered that the conditions assumed in these problems are never realized in practice, for the heat generated by the combustion is never wholly retained in the products. The process of combustion requires a certain time, and during this time a portion of the heat escapes. Moreover, more air passes through the combustible than is required for perfect combustion, and many of the data which enter into the calculation are uncertain. The results, therefore, can only be regarded as approximate. The theoretical conditions are most nearly realized in a gas flame, and especially in that form of burner known as the Bunsen lamp. The temperature of the flame of this lamp, when carefully regulated, is very nearly that which the theory would assign.

63. *Point of Ignition.* — In order that a combustible body should take fire, and continue burning in the atmosphere, it must be heated to a certain temperature, and maintained at this temperature. This temperature is called the point of ignition; and although it cannot always be accurately measured, and is undoubtedly more or less variable under different conditions, yet, nevertheless, it is tolerably constant for each substance. For different substances it differs very greatly. Thus phosphorus takes fire below the boiling point of water, sulphur at $260°$, wood at a low red heat, anthracite coal only at a full red heat, while iron requires the highest temperature of a forge. If a burning body is cooled below its point of ignition, it goes out; and our ordinary combustibles continue burning in the air only because the heat evolved by the burning maintains the temperature above the required point. If the temperature of the combustible is not maintained sufficiently high, either because the chemical union is too slow, or because the calorific power is too small, then the combustible will not continue to burn in the air of itself, although it may burn most readily if its temperature is sustained by artificial means. Hence many of the metals which will not burn in the air burn readily in the flame of a blowpipe, and an iron watch-spring burns like a match in an atmosphere of pure oxygen. The calorific intensity of all combustibles, when burnt in the atmosphere, is, as we have seen, greatly reduced by the presence of nitrogen; and hence it is that, although the burning watch-spring is maintained above the point of ignition in pure oxygen, it soon falls below this temperature, and goes out when ignited in the air. Thus it is that the nitrogen of our atmosphere exerts a most important influence on the action of the fire element; and it can easily be seen that, were it not for these provisions in the constitution of nature, by which the active energies of oxygen are kept within certain limits, no combustible material could exist on the surface of the earth.

64. *Calorific Power derived from the Sun.* — The great mass of the crust of our globe consists of saturated oxygen compounds, or, in other words, of burnt materials; and the total amount of combustible materials which exists on its surface is, comparatively, very small. That which exists naturally consists almost entirely of carbon and its compounds, — such as coal,

naphtha, and wood; and all these substances are the results of vegetable growth, either of the present age or of earlier geological epochs. Moreover, whatever subsequent changes the material may have undergone, it was all originally prepared by the plant from the carbonic acid and water of our atmosphere; for, in the economy of nature, these products of combustion have been made the food of the vegetable world. The sun's rays, acting on the green leaves of the plant, exert a mysterious power, which decomposes carbonic anhydride, and perhaps also water; and, as the result of this process, oxygen is returned to the atmosphere, while carbon and hydrogen are stored up in the growing tissues of the plant. The sun thus undoes the work of combustion, and parts the atoms which the chemical affinities had drawn together. In doing this, the sun exerts an enormous power; and the work which it thus accomplishes is the precise measure of the calorific power of the combustible material, which it then prepares. When we wind up the weight of a clock, we exert a certain power which reappears in its subsequent motions; and so, when the sun's rays part these atoms, the great power it exerts is again called into action, when in the process of combustion the atoms reunite. Moreover, what is true of calorific power is true of all manifestations of power on the surface of the earth. Every form of motion is sustained by the running down of some weight which the sun has wound up; and, according to the best theory we can form, the sun's power itself is sustained by the gradual falling of the whole mass of the solar system towards its common centre. However varying in its manifestation, all power in its essence is the same, and the total amount of power in the universe is constant.

65. *Heat of Chemical Combinations.* — The heat of combustion is only a striking manifestation of a very general principle, which holds true in all chemical changes. It would appear that whenever, in a chemical reaction, atoms or molecules are drawn together by their mutual affinities, a certain amount of moving power is developed, which takes the form of heat; and whenever, on the other hand, these same atoms or molecules are drawn apart by the action of some superior force, the same amount of moving power is expended, and heat disappears. Every chemical reaction is a mixed effect of such combina-

tions and decompositions, and it is simply a complex problem in the mechanical theory of heat to determine what must be in any case the thermal effect. The numerous facts with which we are acquainted in regard to the heat of chemical combination generally agree with the mechanical theory; and, where the facts do not appear to conform to it, the discrepancy probably arises from our ignorance of the nature of the chemical change in question. It would be incompatible with our design to discuss these facts in this book. It must be sufficient to state a few general results, which may be summed up in the following propositions: —

First. The heat absorbed in the decomposition of a compound is equal to the heat evolved in its formation, provided the initial and the final states are the same.

Second. The heat evolved in a series of successive chemical changes is equal to the sum of the quantities which would be evolved in each separately, provided the bodies are finally brought into identical conditions.

Third. The difference between the quantities of heat evolved in two series of changes starting from two different states, but ending in the same final state, is equal to that which is evolved or absorbed in passing from one initial condition to the other.

For example, if a body m evolves a certain amount of heat in uniting with n to form $m\,n$, and if the body $m\,n$ is decomposed by a third body p, so that $m\,p$ is formed, the quantity of heat evolved in this last reaction is less than that which would be evolved in the direct union of m and p by the amount evolved in the formation of $m\,n$.

All these propositions, however, are but special cases under a more general principle which is at the basis of the whole mechanical theory of heat, and which may be enunciated as follows: Whenever a system of bodies undergoes chemical or physical changes, and passes into another condition, whatever may have been the nature or succession of the changes, the quantity of heat evolved or absorbed depends solely on the initial and final conditions of the system, provided no mechanical effect has been produced on bodies outside.

Questions and Problems.

1. How many times more space does the carbonic anhydride formed by burning charcoal ($Sp. Gr. = 2$) occupy than the charcoal burnt?
Ans. One cubic centimetre or two grammes of charcoal yields 3.720 litres. Hence the gas occupies 3.720 times the volume of the charcoal.

2. How many litres of oxygen gas are required to burn one litre of alcohol vapor, and how many litres of aqueous vapor, and how many of carbonic anhydride, will be formed in the process?
Ans. 3 litres of oxygen, 3 litres of aqueous vapor, 2 litres of carbonic anhydride.

3. Given the symbol of alcohol C_2H_6O to find its calorific power.
Ans. 6,572 units, or 7,200 units, assuming that the steam formed was condensed.

4. The composition of dried peat is as follows: Carbon, 625.4; Hydrogen, 68.1; Oxygen, 292.4; Nitrogen, 14.1. Find the calorific power. Ans. 5,521 units.

5. Find the calorific intensity of marsh gas burnt in oxygen.

$$CH_4 + 2O=O = CO_2 + 2H_2O$$

Calorific power of marsh gas, 13,063. Specific heat of steam, 0.4805; of CO_2, 0.2164. Ans. 7,793.

6. Find the calorific intensity of olefiant gas burnt in oxygen.

$$C_2H_4 + 3O=O = 2CO_2 + 2H_2O$$

Calorific power of C_2H_4 11,858. Specific heat of steam and carbonic anhydride as in last problem. Ans. 9,136°.

7. Find the calorific intensity of marsh gas and olefiant gas burnt in air. Besides the data already given, we have also specific heat of nitrogen 0.244. Ans. 2,662°, and 2,916°.

CHAPTER XIII.

MOLECULAR WEIGHT AND CONSTITUTION.

66. *Determination of Molecular Weights.* — It has already been stated that the molecular weight of a substance is an essential element in fixing its symbol and in judging of its chemical relations, but until now the student has not possessed the knowledge necessary in order to understand the methods by which this important constant is determined.

Whenever the substance is a gas, or is capable of being volatilized without decomposition at a manageable temperature, we always ascertain the molecular weight from the specific gravity on the principle already several times enforced (17). The problem then resolves itself into finding the specific gravity of the substance in the state of gas. The methods used in such cases are described on page 21, and more in detail in the author's work on Chemical Physics (330 *et seq.*), and in the same book tables are given which very greatly facilitate the calculation of the results. The specific gravity of the gas or vapor having been found by either of these methods, and referred to hydrogen gas as the unit, the molecular weight of the substance is simply twice the number thus determined. But in applying this important principle, on which our modern chemical philosophy so greatly rests, two precautions are essential.

It is only true that equal volumes of all substances contain the same number of molecules when they are in the condition of true gases. Now, while some substances, like alcohol, assume this condition at temperatures only a few degrees above their boiling point, at least nearly enough for all practical purposes, others, like acetic acid, only attain it at temperatures one or two hundred degrees above their boiling point, and others still, like sulphur, only at the very highest temperatures at which we have been able to experiment. For this reason, the specific gravity of sulphur vapor was for a long time an anomalous fact in the science, and it was not until St. Clair Deville, by using a porcelain globe, succeeded in determining its specific gravity at a very high temperature, that its value was found to correspond with the probable molecular weight, and it is possible that a similar anomaly which still exists in the case of phosphorus and arsenic may be due to the same cause.

The chemist, however, can always have a sure criterion of the condition of any vapor whose specific gravity he is determining by repeating his experiment at a somewhat higher temperature. If the second result does not agree with the first, it is a proof that the vapor is not yet in a proper condition, and that the temperature employed in the experiment was too low. A series of determinations of the specific gravity of the vapor of acetic acid made by Cahours furnish an excellent illustration of the importance of the precaution we are discussing, and will also point out another important relation of this whole subject. This acid when in the most concentrated state boils at 120°, and the specific gravity of its vapor referred to hydrogen at the same temperature and pressure was found to have the following values at the temperatures annexed: —

At 125°	45.90	At 170°	35.30	At 240°	30.16
" 130	44.82	" 180	35.19	" 270	30.14
" 140	41.96	" 190	34.33	" 310	30.10
" 150	39.37	" 200	32.44	" 320	30.07
" 160	37.59	" 220	30.77	" 336	30.07

It will be noticed that, as the temperature increases, the specific gravity diminishes, at first very rapidly, afterwards more slowly, and does not become constant until the temperature has risen 200° above the boiling point, when we have the true specific gravity of acetic acid in the state of gas. This gives for the molecular weight of acetic acid 60 very nearly, which corresponds to the received formula, $C_2H_4O_2$. The slight difference between the theoretical and the observed results may be in part due to errors of observation, but is most probably to be referred to the same cause which determines even in the permanent gases, when under the atmospheric pressure, a variation from Mariotte's law. We do not expect, moreover, to find from the specific gravity the exact molecular weight. *The precise value is determined by the results of analysis, which are, as a rule, far more accurate, and the specific gravity is* only used to decide which of several possible multiples must be the true value. (Compare carefully § 23.)

67. *Disassociation.* — But, besides taking care that the temperature is sufficiently high to bring the substance we are studying into the condition of a true gas, we must look out that the compound is not decomposed in the process. It is now well known that at very high temperatures the disassociation of the elements of a compound body is a constant result, and it is probable that in some cases the same effect is produced at the much lower temperatures which are employed in the determination of vapor densities. The specific gravity of the vapor of ammonic chloride, instead of being 26.75, as we should expect from the undoubted weight of its molecule, NH_4Cl, is only about one half of this amount; and the reason probably is, that, when heated, the molecule breaks into two, and in consequence the volume of the vapor doubles.

$$\boxed{NH_4Cl} = \boxed{NH_3} + \boxed{HCl}$$

It is very difficult, however, to obtain any further evidence that such a change has taken place; for, as soon as the temperature falls, the molecules recombine in assuming the solid condition, and all the phenomena attending the change of state are precisely the same as those observed in any other volatile body. Indeed, although many very ingenious experiments have been made with a view of settling the question, it is still uncertain, not only in this, but also in several other cases, whether disassociation has taken place or not. The question is of great importance to the theory of chemistry. If disassociation does not take place, the cases referred to are exceptions to the law of equal molecular volumes, and specific gravity can no longer be regarded, as now, the sole measure of molecular weight. If, however, it can be proved that such a change does take place, then the unity of our present theory is preserved, and the chemist has only to guard against this cause of error in his experiments.

68. *Indirect Determination of Molecular Weight.* — Although our modern chemical theories rest in great measure on the molecular weight of a few typical compounds determined,

at least approximately, by their specific gravities, yet it is only in a comparatively few cases that we are able to refer the molecular weight of a substance directly to this fundamental measure. Most substances are so fixed, or so easily decomposed by heat, that it is impossible to determine the specific gravity of their vapor, even when such a condition is possible. In these cases, however, we endeavor to refer the molecular weight indirectly to the fundamental measure, by establishing a relation of chemical equivalency between the substance whose molecular weight is sought and some *closely allied* volatile substance whose molecular weight has been previously determined in the manner described above. A few examples will make the application of this principle intelligible.

It is required to determine the molecular weight of nitric acid. A careful study of the numerous nitrates leads to the conclusion that this acid, like hydrochloric acid, HCl, contains but one atom of replaceable hydrogen. For example, we find but one potassic nitrate and one sodic nitrate, whereas we should expect to find several, if the acid were polybasic. Hence we conclude that one molecule of argentic nitrate, like one molecule of argentic chloride, $AgCl$, contains but one atom of silver. Next, we analyze argentic nitrate, and find that 100 parts of the salt contain 63.53 parts of silver. We know the atomic weight of silver, 108, and evidently this must bear the same relation to the molecular weight of argentic nitrate that 63.53 bears to 100. But $63.53 : 100 = 108 : x = 170$, which is the molecular weight of argentic nitrate, and, since the molecule of nitric acid differs from that of argentic nitrate only in containing an atom of hydrogen in place of the atom of silver, its own weight must be $170 - 108 + 1 = 63$.

It is required to determine the molecular weight of sulphuric acid. A comparison of the different sulphates shows that sulphuric acid is dibasic. We find two sulphates of potassium and sodium, an acid sulphate and a neutral sulphate, and hence we conclude that this acid contains two replaceable atoms of hydrogen, and hence that one molecule of neutral potassic sulphate contains two atoms of potassium. In analyzing potassic sulphate it appears that 100 parts of the salt contain 44.83 parts of potassium, and evidently this weight bears the same relation to 100 that the weight of two atoms of potassium bears to the weight of the molecule of potassic sulphate. Thus we have, —

$44.83 : 100 = 78 : x = 174$; the $M.\ W.$ of Potassic Sulphate, and $174 - 78 + 2 = 98$; the $M.\ W.$ of Sulphuric Acid.

By a similar course of reasoning we may deduce from the results of analysis, and from the general chemical relations, the molecular weight of any other acid or base. If there is any question in regard to the basicity of the acid or the acidity of the base, there will be the same question as to the molecular weight; but we cannot be led far into error, for the true weight will be some simple multiple or submultiple of the one assumed, and the progress of science will sooner or later correct our mistake. From the molecular weight of any acid we easily deduce the molecular weights of all its salts.

When the substance is not distinctively an acid or a base, but is capable of entering into combination with other bodies, we can frequently discover its molecular weight by determining experimentally how much of this substance is equivalent to a known weight of some allied but volatile substance whose molecular weight is known. Thus ammonia gas, whose molecular weight is one of the best-established data of chemistry, enters into direct union with a compound of platinic chloride and hydrochloric acid ($PtCl_6H_2$) to form a definite crystalline salt whose composition is exactly known.

$$PtCl_6H_2 + 2NH_3 = PtCl_6(NH_4)_2. \qquad [61]$$

Now a very large number of substances allied to ammonia form with this same platinum salt equally definite products, so that by simply determining the weight of platinum in these compounds, which is very easily done, their molecular weights may at once be referred to the molecular weight of ammonia.

Lastly, if other means fail, we may sometimes discover the molecular weight of a compound by carefully studying the reactions by which it is formed or decomposed, and inferring the weight of the compound from that of its factors or products. We seek to express the reaction in the simplest possible way, and give that value to the molecular weight which best satisfies the chemical equation. Evidently, however, such results are less trustworthy than those obtained by either of the other methods.

69. *Constitution of Molecules.* — It is a favorite theory with some chemists that no molecule can exist in a free condition with any of its affinities unsatisfied, but those who hold this view are compelled to admit that two points of attraction in the same atom may, in certain cases, neutralize each other. Hence, they would distinguish between a dyad atom like that of oxygen (· ·), with its affinities open, and a dyad atom like that of mercury (- -), with its affinities closed through their own mutual attraction. The first could not exist in a free condition, while the last could. In like manner any atom, having an even number of points of attraction, can exist in a free state because all its affinities may be satisfied within itself; but an atom having an uneven number of points cannot, for at least one of its affinities must be open as is shown by the symbol (- - ·). As thus interpreted it must be admitted that the theory explains many facts.

For example, among the univalent elements, chlorine, bromine and iodine are all known to have molecules consisting of two atoms. So, also, the molecule of cyanogen gas consists of two atoms of the radical CN, and the same is true of ethyl, propyl, &c., at least if the hydrocarbons so named have really the constitution first assigned to them.

Passing next to the dyads, we find that, while oxygen, sulphur, selenium and tellurium have molecules consisting of two atoms, the metals mercury and cadmium, and the radicals ethylene, propylene, &c. (C_2H_4 and C_3H_6), have molecules which coincide with their atoms.

Of the well-defined triad elements none are volatile, but the two triad radicals which have been obtained in a free state — allyl[1] (C_3H_5) and kakodyl ((CH_3)$_2As$) — both have double atomic molecules.

In like manner none of the tetrad elements are volatile, and the only tetrad radicals known in a free state have single atomic molecules.

Of the pentad elements nitrogen has a molecule of two atoms, while phosphorus and arsenic have molecules of four

---

[1] See page 78, Problem 7.

atoms. No compound radicals of this order are known in a free state.

Lastly, the only hexad radical known in a free state, benzine, C_6H_6, has a molecule which coincides with its atom.

Thus it appears that in general the theory is sustained by the facts. Nevertheless, there are several well-marked exceptions to it. Thus the well-known compounds $\overset{\text{III}}{NO}$ and $\overset{\text{I}}{NO_2}$ have molecules which act as radicals of uneven atomicities and yet contain but one complex atom. We must be careful, therefore, not to give too much weight to this hypothesis, but still it may be useful in co-ordinating facts. It leads at once to three general principles which will be found to be almost universally true.

The first is that the sum of the atomicities of the atoms of every molecule is an even number.

The second is that the atomicity of any radical is an odd or even number according as the sum of the atomicities of its elementary atoms is odd or even.

The third is that the quantivalence of elementary atoms must be, as stated on page 59, either even or odd. They are *artiads* or *perissads*, and the two characters can never be manifested by the same elements.

It has also been a question among chemists whether molecular combination was possible; in other words, whether it is possible for molecules of different kinds to combine chemically, each preserving its integrity in the compound. Some of the advocates of the unitary theory, in the reaction against the dualistic system, have been inclined to doubt the possibility of such compounds, and have attempted to represent the symbols of *all* compounds in a single molecular group; but any antecedent improbability, on theoretical grounds, is far more than outweighed by the evidence of a large number of compounds whose constitution is most simply explained on the hypothesis of molecular combination. For example, in the crystalline salts it is impossible to doubt that the water exists as such, not as a part of the salt molecule, but combined with it as a whole. So, also, there are a number of double salts whose constitution is most simply explained on a similar hypothesis, and, in the present state of the science, it seems unnecessary to complicate their symbols by forcing them into the unitary mould. It is a characteristic of such molecular compounds as are here assumed, that the force which holds together the molecules is much feebler than that which binds together the atoms in the molecule. When the molecular attraction is very strong, it is probable that in almost all cases the different molecules coalesce into one; and between the extreme limits we find compounds in which it is difficult to determine whether true molecular combination exists or not. Such coalescing of distinct molecules seems always, however, to be attended with a greater development of heat, and, in general, with a more marked manifestation of physical energies, than usually attends either molecular aggregation or atomic metathesis.

In the notation of this book molecular combination is indicated by writing together the symbols of the different molecules thus united, but separating these symbols by periods. Thus the symbols $4KCl.PtCl_4$, and $3NaF.SbF_3$ represent compounds of this class.

70. *Isomerism, Allotropism, Polymorphism.* — We should infer from the doctrine of chemical types that the same atoms might be grouped together in different ways, so as to form different molecules which in their aggregation would present essentially distinct qualities. Hence, we should expect to find distinct substances having the same composition; and in fact our science, organic chemistry especially, is rich in examples of this kind. Such substances are said to be isomeric, and the phenomenon is called isomerism. There are different phases of isomerism, which it will be well to distinguish, not so much on account of any essential differences in the phenomena as in order to make ourselves better acquainted with its manifestations.

In the first place, we have examples of isomeric bodies having the same centesimal composition, but showing no relation to each other in their properties or in their chemical reactions. Sometimes we have assigned to them the same formula, but in other cases the symbol of one is a simple multiple of that of the other. Thus aldehyde and oxide of ethylene have both the symbol C_2H_4O; cane sugar and gum arabic, the common formula $C_{12}H_{22}O_{11}$; lactic acid, the formula $C_3H_6O_3$; and glucose, $C_6H_{12}O_6$. These compounds bear no resemblance to each other, and have no relations in common save the single fact that their centesimal composition is the same.

In the second place, we have numerous examples of isomeric compounds which belong to the same chemical type, and therefore present the same chemical reactions, but in which the two factors of the molecule are in a measure complementary to each other. Thus ethylic formiate has exactly the same composition as methylic acetate,

$$(C_2H_5)\text{-}O\text{-}(CHO), \qquad (CH_3)\text{-}O\text{-}(C_2H_3O);$$

for while the basic radical of the first contains the quantity CH_2 more than the basic radical of the second, the acid radical of the first falls short of the acid radical of the second by exactly the same amount.

In the third place, we have several groups of isomeric compounds, especially among the hydrocarbons, which have the same general properties and the same percentage composition, but which differ from each other in their molecular weights; so that the symbol of one is a multiple of that of the rest. The hydrocarbous ethylene C_2H_4, propylene C_3H_6, butylene C_4H_8, form a group of this kind. Compounds of this class are frequently called *polymeric*, and sometimes the heavier compounds may be regarded as condensed forms of the lighter.

Lastly, we may distinguish still a fourth class of isomeric compounds which have the same general properties, the same symbol, and the same general system of reactions, but which differ in a few marked qualities, physical or chemical, and which preserve these characteristics to a greater or less extent in their compounds. The two forms of toluic acid, $C_8H_8O_2$, belong to this class, and such compounds are isomeric in the fullest sense of the word.

In all the above examples the differences between the isomeric compounds are sufficiently great to lead chemists to assign to each a distinct name. When, however, the differences are not sufficiently great to justify a distinct name, the two bodies are said to be different *allotropic* states of the same substance. Thus there are two varieties of tartaric acid; the first of which deviates the plane of polarization of a ray of light to the left, while the second deviates it to the right; but since in almost every other respect these two bodies are identical, we do not speak of them as different substances, but merely as different allotropic states of tartaric acid. There are also three other varieties of tartaric acid, but these differ so greatly from the normal acid in crystalline form, in solubility, and also in other relations, that they may fairly be regarded as distinct substances.

Again, there are many substances where the difference of state or *allotropism* is associated with difference of crystalline form; and when this difference of form is fundamental, the substance is said to be dimorphous or trimorphous, as the case may be, and the phenomenon is called polymorphism. Thus common calcic carbonate crystallizes in two fundamentally distinct forms, corresponding to the two mineralogical species, calcite and aragonite. Such difference of form, however, is invariably accompanied by a marked difference of properties, so that polymorphism is merely one of the indications of allotropism.

Differences of condition similar to those we have described manifest themselves even more markedly among elementary substances; and indeed the word allotropism was first applied to phenomena of this last class. Thus there are two allotropic states of phosphorus, which differ so much from each other that no one would suspect from their external characters that there was any identity between them, and to these two states correspond two fundamentally different crystalline forms. In some cases the differences between the allotropic states of the same element are far greater than any which are seen between the most unlike isomeric compounds. No substances could be better defined by well-marked and utterly distinct qualities than diamond, plumbago, and charcoal, and yet they are all three allotropic modifications of the one elemental substance we call carbon ; and such phenomena as these give us strong grounds for believing that our present elements may have a composite structure.

Questions and Problems

1. What are the molecular weights of alcohol and camphor as deduced from the results of the Sp. Gr. determinations given on page 23 ?

Ans. 45.5 and 155, which, although not closely agreeing with the theoretical numbers, enables us to decide that the symbols of these compounds are C_2H_6O and $C_{10}H_{16}O$ as the simplest interpretation of the analyses would indicate.

2. At the temperature of 470° the Sp. Gr. of the vapor of sulphuric acid is approximately 1.697. How does this result agree with the generally received symbol of this compound, and how do you explain the discrepancy ?

3. A study of the different tartrates has led to the conclusion already expressed that tartaric acid, although tetratomic, is dibasic. It also appears that one hundred parts of neutral argentic tartrate yield when ignited 55.39 parts of metallic silver. Required the molecular weight of tartaric acid. Ans. 176.

4. An hundred parts of baric oxide, BaO, (whose composition is assumed to be known) yield when treated with sulphuric acid 152.3 parts of baric sulphate. Further it is assumed, as the result of careful study, that sulphuric acid is bibasic, and the metal barium a bivalent radical. Required the molecular weight of sulphuric acid.

Ans. 98.

5. The well-known base aniline gives with platinic chloride a definite crystalline product, one hundred parts of which yield on ignition 32.99 parts of platinum. Required the molecular weight of aniline. How does this result agree with the Sp. Gr. of aniline vapor, which has been found by observation to be 3.210. ?

Ans. 93; which corresponds to Sp. Gr. of 3.223.

6. The base triethylamine gives in like manner a platinum salt, one hundred parts of which yield on ignition 33.67 parts of platinum. Required the molecular weight. Ans. 101.

7. Compare together the symbols of the compounds of the various alcohol radicals on pages 90 to 93 and point out the examples of isomerism.

CHAPTER XIV.

CRYSTALLINE FORMS.

71. *Relations to Chemistry.* — Almost every substance affects a definite polyhedral form, although it may manifest this tendency only under favorable conditions. Such forms are called crystals, and the process of crystalline growth, or development, is called crystallization. The one essential condition of crystallization is a certain freedom of motion, and crystals, more or less perfect, are usually formed whenever a molten liquid "sets," or a solid is deposited from a condition of solution or of vapor; and in each case the slower the process the larger and the more perfect are the crystals. The crystalline condition is, in fact, the normal state of solid matter. It is true that there are a few substances which, like glue, are only known in the colloid state; but in most of the so-called colloid substances this state is abnormal, and there is a constant tendency to crystallization. Moreover, its peculiar crystalline form is one of the most characteristic, and apparently one of the most essential, properties of a substance, and is therefore of great value in determining its chemical affinities. The study of the geometrical relations of these forms is, however, in itself a separate science, and in this connection we can only dwell on the few elementary principles of the subject on which our system of chemical classification in part rests.

72. *Definitions.* — In the forms of crystals the idea of symmetry is the great controlling principle. Each substance follows a certain law of symmetry, which seems to be inherent, and a part of its very nature; and when, from any cause, the character of the symmetry changes, the substance loses its identity, and, even if its chemical composition remains the same, it becomes, to all intents and purposes, a different substance. In every crystal the symmetry points to a few *directions*, to which not only the position of the planes, but also the physical properties of the body, are closely related. Certain of

these *directions*, more or less arbitrarily chosen, are called the *axes* of the crystals, and a *crystalline form* may be defined as a group of *similar* planes symmetrically disposed around these axes. As is evident from this definition a crystalline form, like a geometrical form, is a pure abstraction, and this conception is carefully to be kept distinct from the idea of a crystal, which implies not only a certain form, but also a certain structure. Moreover, in by far the larger number of cases the same crystal is bounded by several forms. Thus, in Fig. 4, which represents a crystal of common quartz, the planes of the prism and the planes of the pyramid are distinct crystalline forms.

Fig. 4.

73. *Systems of Crystals.* — A careful study of the forms of crystals has shown that these forms may be classified under six crystalline systems, each of which is distinguished by a peculiar plan of symmetry. These divisions, it is true, are in a measure arbitrary; for here, as elsewhere in nature, no sharp dividing lines are found; but nevertheless the distinctions on which the classification rests are clearly marked. We can only give in this book a very imperfect idea of these several plans of symmetry by representing with figures a few of the more characteristic forms of each.

74. *First or Isometric System.*[1] — The three most frequently occurring forms of this system are the regular octahedron, the

 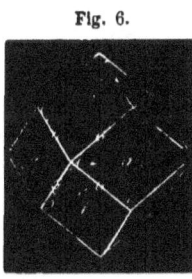 

Fig. 5.   Fig. 6.   Fig. 7.

rhombic dodecahedron and the cube, Figs. 5, 6, and 7. These and all the other forms of the system may be regarded as

[1] Called also monometric.

grouped around three equal and similar axes at right angles to each other, and hence the name isometric (equal dimensions). They present the same symmetry on all sides, and the appearance of the form is identical, whichever axis is placed in a vertical position. In this system no variation in the relative positions or lengths of the axes is possible, for this would change the plan of symmetry on which the system is based.

75. *Second or Tetragonal System.*[1] — The plan of symmetry in this system is best illustrated by the square octahedron, Fig. 8. Of this form the basal section, Fig. 9, is a square, and to

Fig. 8. Fig. 9. Fig. 10.

this fact the name of the system refers. The vertical section, on the other hand, is a rhomb, Fig 10. Here, as in the first system, the forms may all be referred to three rectangular axes, but only two have the same length; the third may be either longer or shorter than the others. The last is the dominant axis of the form, and hence we always place it in a vertical position and call it the vertical axis. The length of the vertical axis bears a constant ratio to that of the lateral axes in all crystals of the same substance, but this ratio differs very greatly for different substances, and is therefore an important crystallographic character. The familiar square prism is another very characteristic form of this system.

Fig. 11. Fig. 12.

Moreover, the planes both of the prism and of the octahedron may have different positions with reference to the lateral axes, as is shown by the two basal sections, Figs. 11 and 12;

[1] Called also dimetric.

and this leads us to distinguish two square prisms and two square octahedrons, one of which is said to be the inverse of the other.

76. *Third or Hexagonal System.* — In the last system the planes were arranged by fours around one dominant axis, while in this system they are arranged by sixes. The most characteristic forms of this system are the hexagonal pyramid, Fig. 13, and the hexagonal prism, Fig. 14. The basal section through either of these forms is a regular hexagon, Fig. 15, and, besides

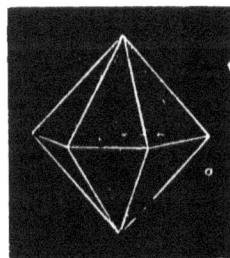

Fig. 13.

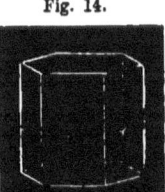

Fig. 14.

Fig. 15.

the dominant or vertical axis, we also distinguish as lateral axes the three diagonals of this hexagonal section. These lateral axes stand at right angles to the vertical axis, but between themselves they subtend angles of 60°. Here, as before, the ratio of the length of the vertical axis to the common length of the lateral axes has a constant value on crystals of the same substance, but differs very greatly with different substances, the vertical axis being sometimes longer and sometimes shorter

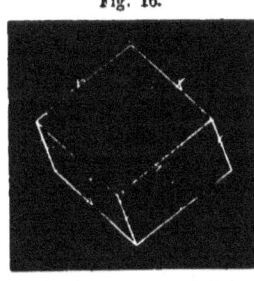

Fig. 16.

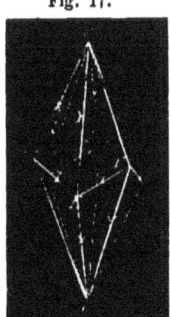

Fig. 17.

than the other three. The rhombohedron, Fig. 16, and the scalenohedron, Fig. 17, are also forms of this system, and occur

CRYSTALLINE FORMS. 141

even more frequently than the more typical forms first mentioned. Lastly, a difference of position in the planes of the prism or pyramid with reference to the lateral axes gives rise in this system to the same distinction between the direct and the inverse forms as in the last.

77. *Fourth or Orthorhombic System.*[1] — The most characteristic forms of this system are the rhombic octahedron, Fig. 18, and the right rhombic prism, from which the system takes its name. The three principal sections of the octahedron, represented by Figs. 19, 20, and 21, and also the basal section of the

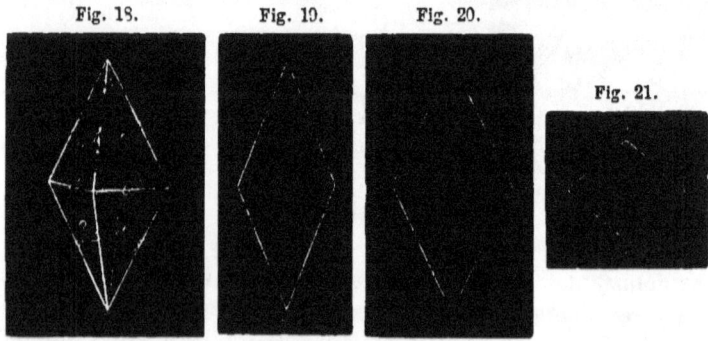

Fig. 18.   Fig. 19.   Fig. 20.   Fig. 21.

prism, are all rhombs, whose relations to the form are indicated by the lettering of the figures. We easily distinguish here three axes at right angles to each other, but of unequal lengths, and in regard to the ratios of these lengths the remarks of the last two sections are strictly applicable.

78. *Fifth or Monoclinic System.* — The forms classed together under this system may be referred to three unequal axes, one of which stands at right angles to the *plane* of the other two, while they are inclined to each other at an angle, which, though constant on crystals of the same substance, varies very greatly with different substances, as vary also the relative dimensions of the axes themselves. Fig. 22 represents an octahedron of this system, and Figs. 23 and 24 represent two sections made through the edges FF and DD' of this form. A section through the edges CC would be similar to Fig. 23, and these three sections give a clear idea of the relative positions of the axes. The section, Fig. 24, containing the two oblique axes,

[1] Called also trimetric.

is called the plane of symmetry, and the faces on all monoclinic crystals are disposed symmetrically solely with reference to this plane. In a word, the symmetry is bilateral, and corresponds

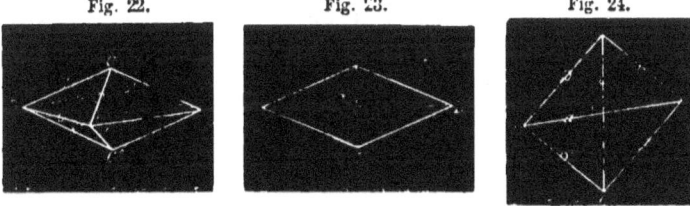

Fig. 22. Fig. 23. Fig. 24.

to the type with which we are so familiar in the structure of the human body. This plan of symmetry is well illustrated by Figs. 25, 26, and 27, which represent the commonly occurring forms of gypsum, augite, and felspar, three of the most common minerals. These figures, however, do not, like those of the previous sections, represent simple crystalline forms. The crystals here represented are in each case bounded by several forms, and indeed in this system such compound forms are alone possible, for no simple monoclinic form can of itself enclose space.

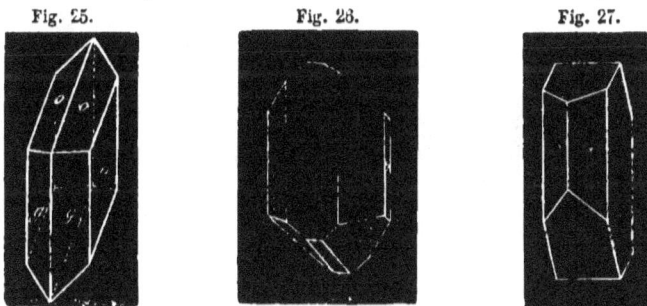

Fig. 25. Fig. 26. Fig. 27.

79. *Sixth or Triclinic System.* — This system is distinguished by an almost complete want of symmetry. Only opposite planes

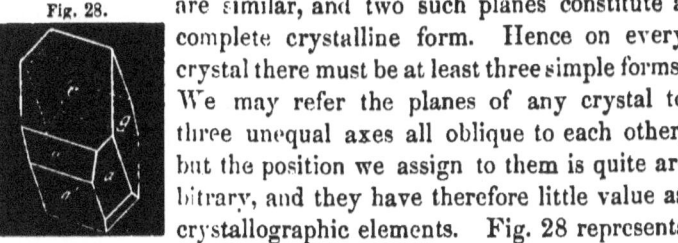

Fig. 28.

are similar, and two such planes constitute a complete crystalline form. Hence on every crystal there must be at least three simple forms. We may refer the planes of any crystal to three unequal axes all oblique to each other, but the position we assign to them is quite arbitrary, and they have therefore little value as crystallographic elements. Fig. 28 represents a crystal of sulphate of copper, one of the very few subtances which crystallize in this system.

CRYSTALLINE FORMS. 143

80. *Modifications on Crystals.* — When several crystalline forms appear on the same crystal, some one is usually more prominent or *dominant* than the rest, and gives to the crystal its general aspect, the planes of the secondary forms only appearing on its edges or solid angles, which are then said to be modified or replaced. Thus, in Figs. 29, 30, and 31, the solid angles of a cube are replaced (or truncated) by the faces of an octahedron; in Fig. 32 the edges of the cube are replaced by the faces of the dodecahedron; in Fig. 33 the edges of the octahedron are modified in the same way; and in Fig. 34 the solid angles of a dodecahedron are replaced by the faces of an

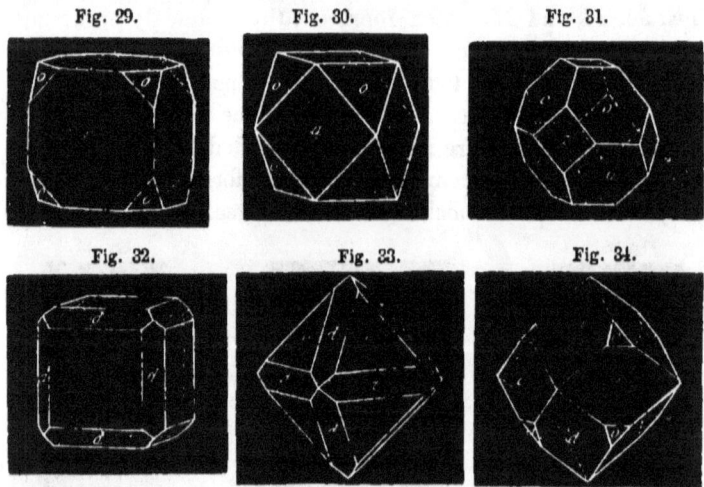

Fig. 29. Fig. 30. Fig. 31.
Fig. 32. Fig. 33. Fig. 34.

octahedron. These are all forms of the isometric system, and the relations of the simple forms to each other, which determine in every case the position of the secondary planes, will be readily seen on comparing together the figures already given on page 138. These figures, like all crystallographic drawings, are geometrical projections, and represent the planes in the same relative position towards the crystalline axes which they have on the crystal itself. Moreover, since in all figures of crystals of this system the axes are drawn in absolutely the same position on the plane of the paper, the same face has also the same position throughout.

As a general rule, *all the similar parts of a crystal are simultaneously and similarly modified.* This important law, which is a simple inference from the principles already stated, is illustrated by the figures just given, and also by Figs.

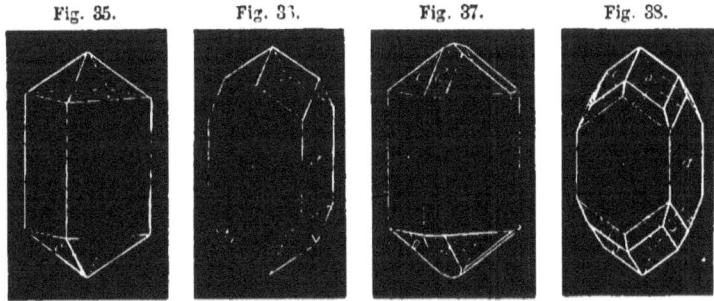

Fig. 35.   Fig. 36.   Fig. 37.   Fig. 38.

35 to 50. By carefully studying the~e figures, as well as Figs. 25 to 28 on page 142, the student will be able to refer each of

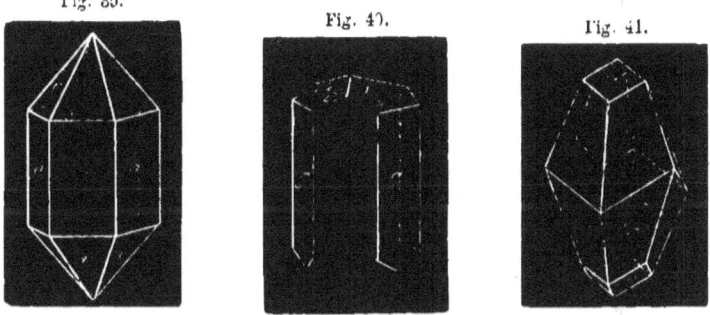

Fig. 39.   Fig. 40.   Fig. 41.

the compound crystals here represented to one or the other of the systems of symmetry already described, and from this and

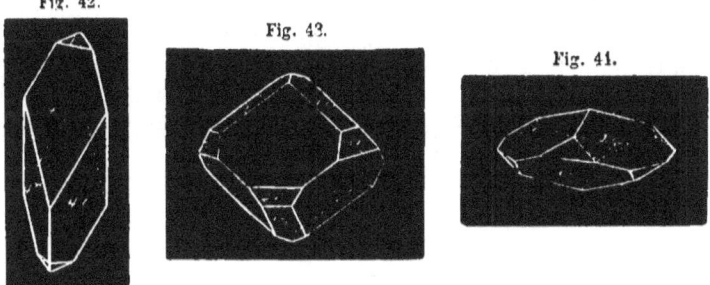

Fig. 42.   Fig. 43.   Fig. 44.

similar practice he will learn, better than from any descriptions, how clearly the modifications on a crystal point out its crystallographic relations.

CRYSTALLINE FORMS. 145

81. Hemihedral Forms. — To the law governing the modifications of crystals just stated, there is one important excep-

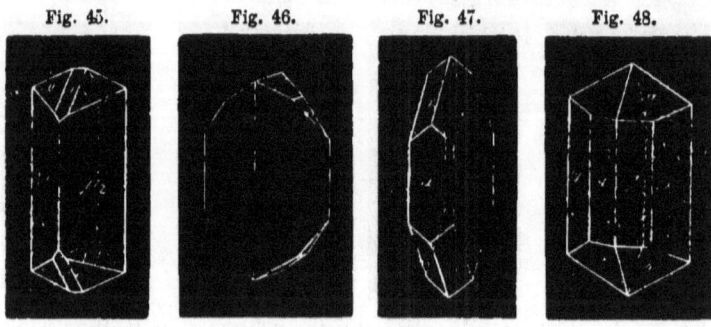

Fig. 45.   Fig. 46.   Fig. 47.   Fig. 48.

tion. It not unfrequently happens that *half the similar parts of a crystal are modified independently of the other half.* Thus

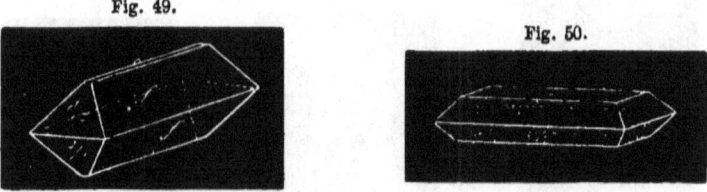

Fig. 49.   Fig. 50.

in Fig. 51 only one half of the solid angles of the cube are truncated. The modifying form in this case is the tetrahedron,

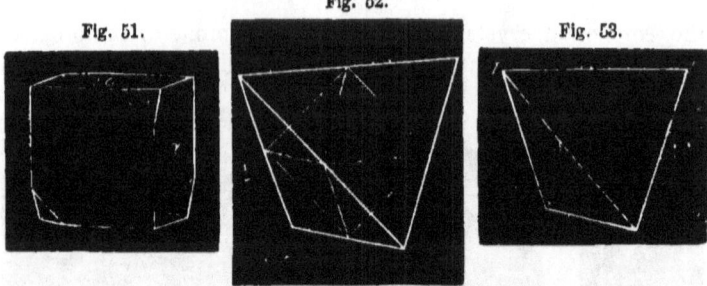

Fig. 51.   Fig. 52.   Fig. 53.

Fig. 53, also a simple form of the isometric system. When all the solid angles of the cube are truncated, the modifying form, as has been shown, is the octahedron, and the relation which the tetrahedron bears to the octahedron is shown by Fig. 52. The rhombohedron, Fig. 54, stands in a similar relation to the hexagonal pyramid, Fig. 55. From these figures it is evident that while the octahedron and the hexagonal pyramid have all the planes which perfect symmetry requires, the

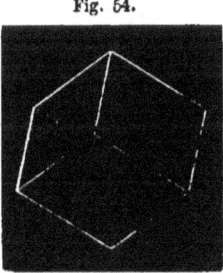
Fig. 54.

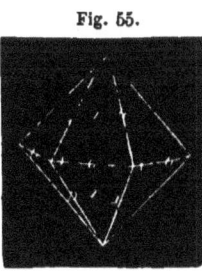
Fig. 55.

tetrahedron and the rhombohedron have only half the number, and in crystallography all forms which bear a similar relation to the forms of perfect symmetry are said to be *hemi*hedral, while the forms of perfect symmetry are distinguished as *holo*hedral. The hemihedral forms are quite numerous in all the systems, but with the exception of the tetrahedron, rhombohedron, and scalenohedron (Fig. 17), they seldom appear except as modifying planes on the edges or solid angles of the more perfect forms. As a general rule, they are easily recognized, but not unfrequently they give to a crystal the aspect of a different system from that to which it really belongs, and may lead to false inferences; but these can, in most cases, be corrected by a careful study of the interfacial angles.

82. *Identity of Crystalline Form.* — As has already been stated, every substance is marked by certain peculiarities of outward form, which are among its most essential qualities, and we must next learn in what these peculiarities consist. As a general rule, the same substance crystallizes in the same form, but under unusual circumstances it frequently appears in other forms of the same system. Thus fluorspar is usually found crystallized in cubes, but in large collections crystals of this mineral may be seen in almost all the holohedral forms of the isometric system, including their numerous combinations. In like manner common salt usually crystallizes in cubes, but out of a solution containing urea it frequently crystallizes in octahedrons. Moreover, the same principle holds true in regard to substances crystallizing in other systems, most of whose forms never appear except in combination. Thus the mineral

quartz generally shows the simple combination represented in Fig. 4; but more than one hundred other forms, all, however, belonging to the same system, have been observed on crystals of this well-known substance. So also the crystals of gypsum, augite, and felspar, in most cases present the forms already figured on page 142, although other forms are common, which, however, in each case all belong to the same crystalline system. We never find the same substance in the forms of different systems except in those cases of polymorphism already described, page 135, where the differences in other properties are so great that the bodies can no longer be regarded as the same substance.

Among substances crystallizing in the isometric system the crystalline form is not so distinctive a character as it is in other cases. In this system the relative dimensions are invariable, and the octahedron, the dodecahedron, and the cube, more or less modified by different replacements, are the constantly recurring forms. Even here, however, specific differences may at times be found in the fact that some substances affect hemihedral forms on modification, while others do not. In all the other systems the dimensions of the crystal (the relative lengths of its axes and the values of the interaxial angles) distinguish each substance from every other. But here, also, the general statement must be somewhat modified.

We frequently find on the crystals of the same substance several forms having different axial dimensions. Thus, on the crystal represented by Fig. 56, belonging to the tetragonal system, there are three different octahedrons, and three corresponding values of the vertical axis. But if, beginning with the planes of the octahedron O, we determine the ratio which its vertical axis bears to the common length of the two lateral axes, and call this value a, we shall find that the corresponding values for the two other octahedrons are $2a$ and $\tfrac{1}{2}a$ respectively. Moreover, if we extend our study we shall also find that this example illustrates a general principle, and that *the crystalline forms of a given substance include not only those of identical axial dimensions, but also those whose dimensions bear to each other some simple ratio.*

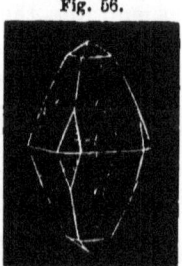

Fig. 56.

This most important law gives to the science of crystallography a mathematical basis, and enables us to apply the exhaustive methods of analytical geometry in discussing the various relations of the subject. Among the actual forms of a given substance we fix on some one as the fundamental form, and, taking the values of its axial dimensions as our standards, we are able to express the position of the planes of all the possible forms by means of very simple symbols, and also to express by mathematical formulæ the relations of the interfacial angles to the same fundamental elements of the crystal; so that the one may readily be calculated from the other.

It may seem at first sight that the crystallographic distinction between different substances, insisted on above, is greatly obscured by the important limitations just made. But it is not so, at least to any great extent. The selection of the fundamental form of a given substance is not arbitrary, although it is based on considerations which it lies beyond the scope of this book to discuss. Moreover, an error in this choice is not fundamental, since the true conception of the form of a substance includes not only the fundamental form, but all those which are related to it. This conception, though not readily embodied in ordinary language, is easily expressed by a general mathematical formula, and is as tangible to one familiar with the subject as the general statement first made.

But however obscure, to those who are not familiar with mathematical conceptions, may be the distinction between the forms of different substances in the same system, the difference between the different systems is clear and definite, and it is with this broad distinction that we have chiefly to deal in our chemical classification.

83. *Irregularities of Crystals.* — It must not be supposed that natural crystals have the same perfection of form and regularity of outline which our figures might seem to indicate. In addition to being more or less bruised or broken from accidental causes, crystals are rarely terminated on all sides, — one or more of the faces being obliterated where the crystal is implanted on the rock, or where it is merged in other crystals. But by far the most remarkable phase which the irregularities of crystals present is that shown by Figs. 57 to 67. By comparing together the figures which have been here grouped to-

CRYSTALLINE FORMS. 149

gether on the page, and which represent in each case different phases of the same crystalline form, it will be seen that the variations from the normal type are caused by the undue de-

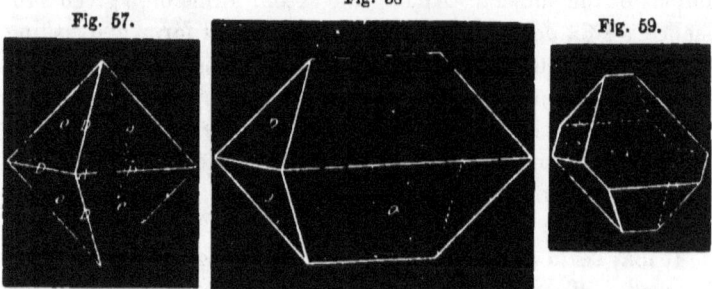

Fig. 57. Fig. 58. Fig. 59.

velopment of certain planes at the expense of their neighbors, or by an abnormal growth of the crystal in some one direction.

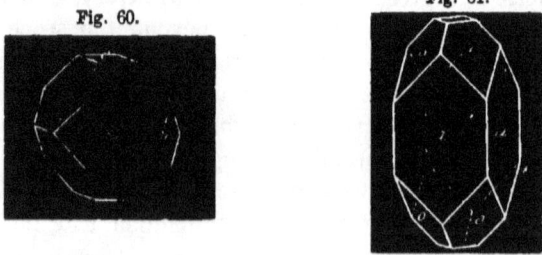

Fig. 60. Fig. 61.

Such forms as these, however, although great departures from the ideal geometrical types, are in perfect harmony with

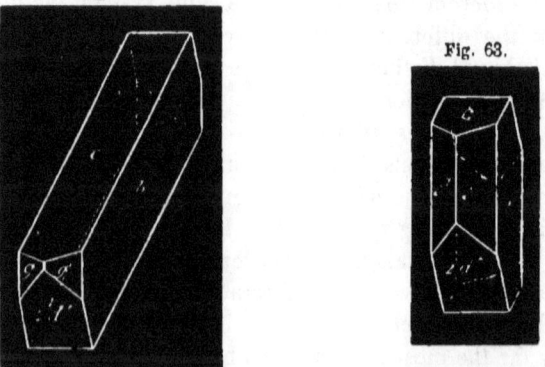

Fig. 62. Fig. 63.

the principles of crystallography. The axis of a crystal is not a definite line, but a definite direction; and the face of a crystal is not a plane of definite size, but simply an extension in two definite directions. These directions are the only fundamental elements of a crystalline form, and they are preserved under

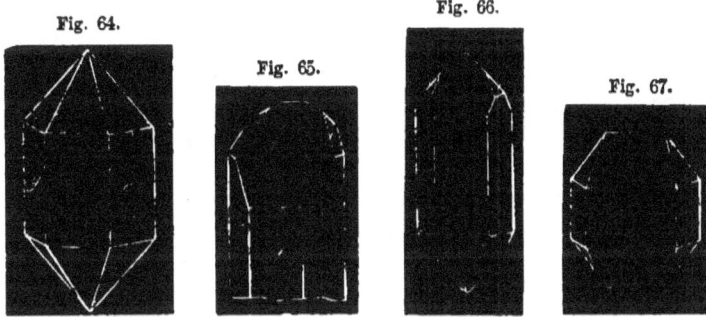

Fig. 64. Fig. 65. Fig. 66. Fig. 67.

all conditions, as is proved by the constancy of the interfacial angles, and of the modifications, on crystals of the same substance, however irregular may have been the development.

84. *Twin Crystals.* — Every crystal appears to grow by the slow accretion of material around some nucleus, which is usually a molecule or a group of molecules of the same substance, and which we may call the crystalline molecule or germ. Now we must suppose that these molecules have the same differences on different sides which we see in the fully developed crystal, and which, for the want of a better term, we may call polarity. As a general rule, in the aggregation of the molecules a perfect parallelism of all the similar parts is preserved. But, if molecular polarity at all resembles magnetic polarity, it may well be that two crystalline molecules might become attached to each other in a reversed position, or in some other definite position determined by the action of the polar forces. Assume now that each of these crystalline molecules " germinates," and the result would be such twin crystals as we actually find in nature. The result is usually the same as if a crystal of the normal form were cut in two by a plane having a definite position towards the crystalline axes, and one part turned half round on the other; and twins of this kind are therefore called hemitropes. Figs. 68 to 71. At other times the germinal molecules seem to have become attached with their dominant axes at right angles to each other, and then there result twins such as are represented in Figs. 72 and 73; and many other modes of twin-

ning are possible. Some substances are much more prone to the formation of twin crystals than others, and the same substance generally affects the same mode of twinning, which may

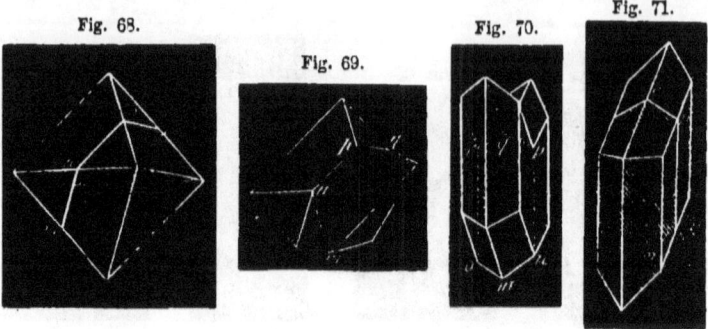

Fig. 68. Fig. 69. Fig. 70. Fig. 71.

thus become an important specific character. The plane which separates the two members of a twin crystal, called the plane

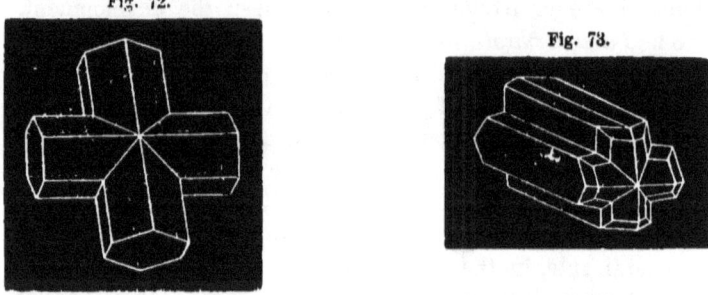

Fig. 72. Fig. 73.

of twinning, has always a definite position, and is in every case parallel either to an actual or to a possible face on both of the two forms.

Twin crystals always preserve the same symmetry of grouping, and the values of the interfacial angles between the two forms are constant on crystals of the same substance, so that they might sometimes be mistaken for simple crystals by unpractised observers. There is, however, a simple criterion by which they can be generally distinguished. Simple crystals never have re-entering angles, and, whenever these occur, the faces which subtend them must belong to two individuals.

The same principle which leads to the formation of twin crystals may determine the grouping of several germinal molecules, and lead to the formation of far more complex combinations. Frequently, as it would seem, a large number of molecules arrange themselves in a line with their principal axes parallel and their dissimilar ends together, and hence result linear groups of crystals alternating in position, but so fused into each other as to leave no evidence of the composite character except the re-entering angles, and frequently these are marked only by the striations on the surface of the resulting faces. Such a structure is peculiar to certain minerals, and the resulting striation frequently serves as an important means of distinction. The orthoclase and the klinoclase felspars are distinguished in this way.

85. *Crystalline Structure.* — The crystalline form of a body is only one of the manifestations of its crystalline structure. This also appears in various physical properties, which are frequently of great value in fixing the crystallographic relations of a substance, and such is especially the case when, on account of the imperfection of the crystals, the crystalline form is obscure. Of these physical qualities one of the most important is *cleavage*.

As a general rule, crystallized bodies may be split more or less readily in certain definite directions, called planes of cleavage, which are always parallel either to an actual or to a possible face on the crystals of the substance, and are thus intimately associated with its crystalline structure. At times the cleavage is very easily obtained, when it is said to be *eminent*, as in the case of mica or gypsum, which can readily be split into exceedingly thin leaves, while in other cases it can only be effected by using some sharp tool and applying considerable mechanical force. With a few unimportant exceptions the cleavage planes have the same position on all specimens of the same substance. Thus specimens of fluor-spar may be readily cleaved parallel to the faces of an octahedron, Fig. 5, those of galena parallel to the faces of a cube, Fig. 7, those of blende parallel to the faces of a dodecahedron, Fig. 6, and those of calc-spar parallel to the faces of a rhombohedron, Fig. 16. In these cases, and in many others, the cleavage is a more distinctive character than the external form, and can be more frequently observed, and we generally regard the form produced by the union of the several planes of cleavage as the fundamental form of the substance.

Again, we always find that cleavage is obtained with equal

ease or difficulty parallel to similar faces, and with unequal ease or difficulty parallel to dissimilar faces. Moreover, the dissimilar cleavage faces thus obtained may generally be distinguished from each other by differences of lustre, striation, and other physical character; and such distinctions are frequently a great help in studying the crystallographic relations of a substance. Similar differences on the natural faces of crystals are also equally valuable guides.

But, of all the modes of investigating the crystalline structure of a body, none can compare in efficiency with the use of polarized light. It is impossible to explain the theory of this beautiful application of the principles of optics without extending this chapter to a length wholly incompatible with the design of this book. It must suffice to say, that if we examine with a polarizing microscope a thin slice of any transparent crystal of either the second or third system, cut parallel to the dominant axis, we see a series of colored rings, intersected by a black cross, and it is evident that the circular form of the rings answers to the perfect symmetry which exists in these systems around the vertical axis. If, however, we examine in a similar way a slice from a crystal of one of the last three systems, cut in a definite direction, which depends on the molecular structure, and must be found by trial, we see a series of oval rings with two distinct centres, indicating that the symmetry is of a different type. Moreover, the distribution of the colors around the two centres corresponds in each case to the peculiarities of the molecular structure, and enables us to decide to which of the three systems the crystal belongs.

The use of polarized light has revealed remarkable differences of structure in different crystals of the same substance, connected with the hemihedral modifications described above. The Figures 74 and 76 represent crystals of two varieties of tartaric acid, which only differ from each other in the position of two hemihedral planes, and are so related that when placed before a mirror the image of one will be the exact representation of the other. The intermediate Figure, 75, represents the same crystal without these modifications. Since the solid angles are all similar, we should expect to find them all modified simultaneously; but, while on crystals of common tartaric acid only the two front angles (as the figure is drawn) are replaced, a variety of this acid has been discovered having similar crystals, whose back angles only are modified. Now, it is found that a solution of the common acid rotates the plane of polarization of a beam of light to the right, while a similar so-

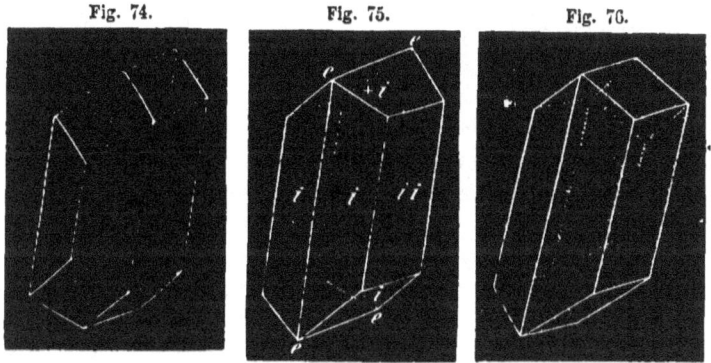

Fig. 74.   Fig. 75.   Fig. 76.

lution of this remarkable variety rotates the plane of polarization to the left. This difference of crystalline structure, moreover, is associated with certain small differences in the chemical qualities of the two bodies; but the difference is so slight that we cannot but regard them as essentially the same substance, and the polarized light thus reveals to us the beginnings of a difference of structure, which, when more developed, manifests itself in the phenomena of isomerism. It is a remarkable fact, worthy of notice in this connection, that these two varieties of tartaric acid chemically combine with each other, forming a new substance called racemic acid.

Questions.

1. By what peculiar mode of symmetry may each of the six crystalline systems be distinguished? How may crystals belonging to the 1st system be recognized? How may crystals of the 2d, 3d, and 4th systems be distinguished by studying the distribution of the similar planes around their terminations or dominant axes? By what peculiar distribution of similar planes may the crystals of the 5th and 6th systems be distinguished from all others? State the system to which each of the crystals, represented by the various figures of this chapter, belongs, and give the reason of your answer in every case.

2. We find in the mineral kingdom two different octahedral forms of titanic acid belonging to the tetragonal system. In one of these forms the ratio of the unequal axes is $1 : 0.6442$, in the other it is $1 : 1.7723$. Can these forms belong to the same mineral substance?

CHAPTER XV.

ELECTRICAL RELATIONS OF THE ATOMS.

• 86. *General Principles.* — If in a vessel of dilute sulphuric acid (one part of acid to twenty of water) we suspend a plate of zinc and a plate of platinum, opposite to each other, and not in contact, we find that no chemical action whatever takes place, provided the zinc and the acid are perfectly pure. As soon, however, as the two plates are united by a copper wire, as represented in Fig. 77, chemical action immediately ensues, and the following phenomena may be observed. First: Bubbles of hydrogen gas are evolved from the surface of the platinum plate. Secondly: The zinc plate slowly dissolves, the zinc combining with the radical of the acid to form zincic sulphate, which is soluble in water. Lastly: A peculiar mode of atomic motion called electricity is transmitted through the copper wire, as may be made evident by appropriate means. If the connection between the plates is broken by dividing the conducting wire, the chemical action instantly stops, and the current of electricity ceases to flow; but, as soon as the connection is renewed, these phenomena again appear.

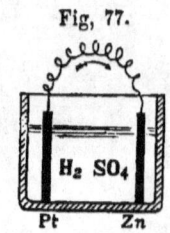

Fig. 77.

Similar effects may be produced by other combinations than the one just mentioned, provided only certain conditions are realized. In the first place, the two plates must consist of materials which are unequally affected by the liquid contained in the vessel, or cell; and the greater the difference in this respect, within manageable limits, the better. In the second place, the materials, both of plates and connector, must be conductors of electricity; and, lastly, the liquid must contain some substance for one of whose radicals the material of one of the plates has sufficient affinity to determine the decomposition of the compound in solution.

Practically, the combination first mentioned, with a few slight modifications, is found to be the best adapted for general use; but, in order to bring the phenomena before our minds in their simplest form, we will assume — other things being the same as before — that the compound in solution is hydrochloric acid, $\overset{+\,-}{HCl}$, since this consists of a simple negative radical united to a simple positive radical. In this case the space between the plates is filled with molecules consisting of hydrogen and chlorine atoms, as is indicated in Fig. 78, where we have attempted to represent by symbols a single one of the innumerable lines of molecules of which we may conceive as uniting the two plates. The zinc plate, in virtue of the powerful affinity of zinc for chlorine, attracts the chlorine atoms, which rush towards it with immense velocity; and the sudden arrest of motion which attends the union of the chlorine with the zinc has the effect of an incessant volley of atomic shot against the face of the plate. Each of the atomic blows must give an impulse to the molecules of the metal itself, which will be transmitted from molecule to molecule, through the material of the plate and the connecting wire, in the same way that a shock is transmitted along a line of ivory balls; and an electric current, as we conceive of it, is merely a wire, or other conductor, filled with innumerable lines of oscillating molecules.

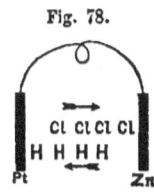

Fig. 78.

But these very impulses, which impart motion to the metallic molecules, react on the liquid, forcing back the hydrogen atoms towards the platinum, and the result is a constant metathesis along the whole line of molecules between the two plates; so that, for every atom of chlorine which enters into union with the zinc, an atom of hydrogen is set free at the face of the platinum plate. Thus we have the singular phenomenon produced of two coexisting atomic currents throughout the mass of the liquids, a stream of chlorine atoms constantly setting towards the zinc plate, and a stream of hydrogen atoms flowing in the opposite direction, in the same space, towards the platinum plate. Corresponding to this motion in the mass of the liquid is the peculiar atomic motion in the metallic conductors. The two, for some unknown reason, are mutually dependent; and the moment the connection is broken, so that the motion can no longer be transmitted through the conductor, the motion in the liquid itself ceases. As regards the mode of atomic motion in the solid metallic conductors, we have been unable to form any clear conceptions. Although apparently allied to heat, this peculiar mode of atomic motion, called electricity, is capable of producing very different classes of effects, and has the remarkable power of imparting to the unlike atoms of almost all compound bodies the same opposite motions which attend its first production. In our ignorance of its nature, the direction we assign to the electric current is in great measure arbitrary; and it is more probable that a twofold current coexists in the conducting wire, corresponding to that which we have recognized as actually flowing through the liquid between the plates of the cell. These two currents have in fact been distinguished by different names; that flowing into the conducting wire from the platinum, or *inactive plate*, being called the positive current, and that from the zinc, or *active plate*, the *negative current*. These names, however, are intended to indicate merely some unknown opposition of relations between the two lines of moving atoms, and not an essential difference in the mode of the motion. Reasoning from certain mechanical phenomena, the physicists originally assumed that the electrical current flowed in but one direction, that is, through the conducting wire from the platinum plate to the zinc, and from the zinc plate through the liquid back again to the platinum; and now, when the direction of the current is spoken of, it is this direction, that of the positive current, which is always meant.

87. *Electrical Conducting Power or Resistance.*— Different materials transmit the electric current with very different degrees of facility; for while in some this peculiar form of molecular motion is easily maintained, in others the molecules yield to it only with difficulty, and many substances seem not to be susceptible of it. The conducting powers of different metallic wires have been very carefully studied, and some of the most trustworthy results are collected in the following table. Silver is the best conductor known, and, assuming that a silver wire of definite size and 100 centimetres long is taken as the standard, the number opposite the name of each metal is the length in centimetres of a wire made of this metal, and of the same size

as the first, which will oppose the same resistance to the transmission of the current. The second column gives the relative resistances of wires of the same materials when of equal size and of equal lengths. The relative or *specific resistances* of two such wires must evidently be inversely proportional to their conducting powers, and thus the numbers of the second column are easily calculated from those of the first. For the results collated in this table we are indebted to the careful investigations of Professor Matthiessen.

Pure Metals.		Conducting Power.		Specific Resistance.	
		At 0°.	At 100°.	At 0°.	At 100°.
Silver	(*hard drawn*)	100.00	71.56	1.000	1.397
Copper	(*hard drawn*)	99.95	70.27	1.0005	1.423
Gold	(*hard drawn*)	77.96	55.90	1.283	1.788
Zinc		29.02	20.67	3.445	4.838
Cadmium		23.72	16.77	4.216	5.964
Cobalt		17.22		5.808	
Iron	(*hard drawn*)	16.81		5.948	
Nickel		13.11		7.628	
Tin		12.36	8.67	8.091	11.53
Thallium		9.16		10.92	
Lead		8.32	5.86	12.02	17.06
Arsenic		4.76	3.33	21.01	30.03
Antimony		4.62	3.26	21.65	30.68
Bismuth		1.245	0.878	80.34	113.9

Commercial Metals.	C. P.	Sp. R.	C°.	Commercial Metals.	C. P.	Sp. R.	C°.
Copper	77.43	1.291	18.8	Iron	14.44	6.924	20.4
Sodium	37.43	2.672	21.7	Palladium	12.64	7.911	17.2
Aluminum	33.76	2.962	19.5	Platinum	10.53	9.497	20.7
Magnesium	25.47	3.926	17.0	Strontium	6.71	14.90	20.5
Calcium	22.14	4.516	16.8	Mercury	1.63	61.35	22.8
Potassium	20.85	4.795	20.4	Tellurium	0.00077	129,800	19.6
Lithium	19.00	5.262	20.0	Red Phosphorus	0.00000123	81,300,000	24.0

If, next, we compare wires of the same material, but of different sizes, we find that the resistance increases as the length, and diminishes as the area, of the section. Moreover, if we adopt some absolute standard of resistance, like that selected by the English physicists, we can easily express the resistance of any given conductor in terms of this unit. It must be remembered, however, in making such comparisons, that the resistance varies with the temperature, and also that the conducting power of the same metal is materially influenced both by its physical condition and by the presence of impurities.

88. *Ohm's Law.* — The first effect of the chemical forces in the cell of an electrical combination is to marshal the dissimilar atoms of the active liquid between the plates into lines, which at once begin to move in parallel columns, but in opposite directions (Fig. 78). Moreover, each one of these lines of *moving* atoms is continued by a corresponding line of *oscillating* atoms in the conducting wire, and thus is formed a continuous circuit returning upon itself. The union of all the *lines of force* in the two opposite coexisting streams constitutes in any case the electrical current, and the different parts of this continuous chain are so related that *the total amount of motion is always the same at every point on the circuit,* and *no more lines of moving atoms form in the liquid between the plates than can be continued through the oscillating atoms of the solid conductors.*

If we adopt this theory, it is obvious that the strength of any electrical current must depend, — first, on the number of continuous lines of force, and secondly, on the strength of the atomic blows transmitted through each of these channels. Of these two elements, the first is determined solely by the total resistance which the various parts of the circuit oppose to the electrical motion, and the greater this resistance the less will be the number of the lines of force. The second element is determined by the value of the resultants of all the chemical forces acting in any combination, which impel the dissimilar atoms towards the opposite plates, — a value which depends solely on *the chemical relations of the materials of the plates to that of the active liquid,* and is what is called the *electromotive force* of the combination, a quantity we will represent by E.

It appears, then, from the above analysis, that an electrical current is a continuous chain, which is sustained in a regulated and equable motion in all its parts by the chemical activity in the cell, and that the strength of this current at any point of the chain must be directly proportional to the electromotive force, and inversely proportional to the sum of the resistances throughout the circuit. If, then, we represent the resistance in the conducting wire by r, the resistance of the liquid between the plates of the cell by R,[1] and also the strength of the current by C, we shall have, in every case,

$$C = \frac{E}{R+r} \qquad [62]$$

[1] The resistance of any circuit may be conveniently divided into two parts,

The quantities C, R, r, and E may all be accurately measured, and stand in each case for a certain number of arbitrary units, whose relations will hereafter be stated.

89. *Electromotive Force and Strength of Current.* — It would seem at first sight as if the strength of an electric current might be increased by simply enlarging the size of the plates in the combination employed, and obviously the number of *possible* lines of moving atoms which could be marshalled in the liquid between the plates would thus be increased; but, as has been stated, the parts of the circuit are so intimately connected that no greater number of lines of atoms can form between the plates than can be continued through the whole circuit, and practically there may be formed between the smallest plates a vastly greater number of atomic lines than can be continued through any conductor, however good its quality or however ample its size. Hence it is, that by increasing the size of the plates we multiply the lines of force only in so far as we thereby lessen the resistance in the liquid part of the circuit. We thus simply lessen the value of R in Ohm's formula [62]; but if this value is already small as compared with r, that is, if the resistance in the cell is small compared with that in the conductor, no material gain in the power of the current, or in the value of C, will result. On the other hand, if the exterior resistance, r, is small, or nearly nothing, as when the plates are connected by a thick metallic conductor, then the value of C will increase in very nearly the same proportion as the size of the plates is enlarged, and the value of R, in consequence, diminished. Under these conditions, the number of lines of moving atoms is greatly multiplied, and we obtain a current of very great volume, but only flowing with the limited force which the single cell is capable of maintaining. Such a current has but little power of overcoming obstacles; and if we attempt to condense it by using a smaller conductor, we reduce, as has been said, the chemical action which keeps the whole in motion, and thus lessen the volume of the flow. This is generally expressed by saying

first, the resistance of the conducting wire, and secondly, the resistance of the liquid portion of the circuit between the two plates of the cell. The resistance of the solid conductor may be readily estimated on the principles stated in the last section, and the resistance of liquid may be measured in a similar way. The last depends, — 1. On the conducting power of the liquid; 2. On the length of the liquid circuit, which is determined by the distance apart of the plates; 3. On the area of the section of the liquid conductor, which is determined by the size of the plates; and, 4. On the temperature.

that the current has large *quantity*, but small *intensity*, or more properly, *electromotive power*.

It must now be obvious from the theory, that we cannot increase effectively the intensity of a current (its power of overcoming obstacles) without in some way increasing the chemical activity, or, in other words, the electro-motive force of the combination employed, and Ohm's formula leads to the same result. If the value of r in our formula is very large as compared with R, we cannot increase it still farther without lessening the total value, C, unless at the same time we increase the value of E. Now, this electro-motive force may be, to a certain extent, increased by using a more active combination; but the limit in this direction is soon reached, and the construction of the cell which has been found practically to be the most efficient will be described below.

We can, however, increase the effective electro-motive force to almost any extent by using a number of cells, and coupling them together in the manner represented by Fig. 79, the platinum plate of the first cell being united by a large metallic connector to the zinc plate of the second, and so on through the line, until finally the external conductor establishes a connection between the platinum plate of the last cell and the zinc plate of the first. Such a combination as this is called a Galvanic or Voltaic[1] battery, and the current which flows through such a combination has a vastly greater power of overcoming resistance than that of any single cell, however large.

The increased effect obtained with such a combination will be easily understood, when it is remembered that each of the innumerable closed chains of moving molecules, now extends through the whole combination, and that all its parts move in the same close mutual dependence as before.

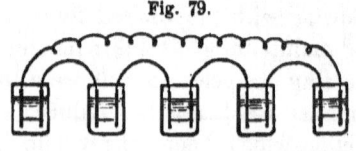

Fig. 79.

But whereas with a single cell the motion throughout any single chain of molecules is sustained by the chemical energy at only one point, it is here reinforced at several points;

---

[1] From the names of Galvani and Volta, two Italian physicists, who first investigated this class of phenomena.

and where before we had a single atomic blow, we have now a number, which simultaneously send their united energy along one and the same line. The effective electro-motive power is then increased in proportion to the number of cells; and the effect on the current would be increased in the same proportion, were it not for the fact that the current must keep in motion a greater mass of liquid, and hence the resistance is increased at the same time. The value of this resistance, however, is easily estimated, since it is directly proportional to the distance through which the current has to flow in the liquid; and hence, if the liquid is the same in all the cells, and the plates are at the same distance apart in each, the liquid resistance will be n times as great in a combination of n cells as it is in one. Moreover, since the effective electro-motive force is n times as great also, while the external resistance remains unchanged, the strength of the current from such a combination will still be expressed by formula [62] slightly modified.

$$C = \frac{nE}{nR + r} \qquad [63]$$

This formula shows at once, that, when the exterior resistance is very small, or nothing, very little or no gain will result from increasing the number of cells, for the ratio of nE to nR is the same as that of E to R; and, under such conditions, in order to increase the strength of the current, we must increase the surface of the plates. If, on the contrary, the exterior resistance is very large, the formula shows that great gain will result from increasing the number of the cells, and that little or no advantage will accrue from enlarging the surface of the plates. Moreover, the formula enables us in any case to determine what proportion the number of cells should bear to the size of the plates in order to obtain the full effect of any battery in doing a given work; and in the numerous applications of electricity in the arts we find abundant illustrations of the principles it involves. The methods used in finding the values of the quantities represented in the formula lie beyond the scope of this work, and for such information the student is referred to works on Physics.

90. *Constructions of Cells.* — It is found practically that the

simple combination of plates and acid first described must be slightly modified in order to obtain the best results.

In the first place, both the zinc and sulphuric acid of commerce contain impurities, which give rise to what is called local action, and cause the zinc to dissolve in the acid when the battery is not in action. Fortunately, however, it has been found that such local action can be wholly prevented by carefully amalgamating the surface of the zinc and filtering the acidulated water.

The mercury on the surface of the zinc plates acts as a solvent, and gives a certain freedom of motion to the particles of the metal. These, by the action of the chemical forces, are brought to the surface of the plate, while the impurities are forced back towards the interior, so that the plate constantly exposes a surface of pure zinc to the action of the acid.

By filtering we remove the particles of plumbic sulphate which remain floating in the sulphuric acid for a long time after it has been diluted with water, and which, when deposited on the surface of the zinc, become points of local action, even when the plates have been carefully amalgamated.

In the second place, the continued action of the simple combination first described develops conditions which soon greatly impair, and at last wholly destroy, its efficiency.

The hydrogen gas, which by the action of the current is evolved at the platinum plate, adheres strongly to its surface, and with its powerful affinities draws back the lines of atoms moving towards the zinc plate, and thus diminishes the effective electro-motive force. Moreover, after the battery has been working for some time, the water becomes charged with zincic sulphate; and then the zinc, following the course of the hydrogen, is also deposited on the surface of the platinum, which after a while becomes, to all intents and purposes, a second zinc plate, and then, of course, the electric current ceases.

Both of these difficulties, however, have also been surmounted by a very simple means discovered by Mr. Grove, of London. The Grove cell, Fig. 80, consists of a circular plate of zinc well amalgamated on its surface, and immersed in a glass jar containing dilute sulphuric acid. Within the zinc cylinder is placed a cylindrical vessel of much smaller diameter, made of porous earthenware, and filled with the strongest nitric acid, and in this hangs the plate of platinum, Fig. 81. The walls of

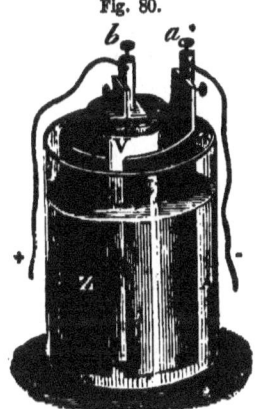

Fig. 80.

Fig. 81.

the porous cell allow both the hydrogen and the zinc atoms to pass freely on their way to the platinum plate; but the moment they reach the nitric acid they are at once oxidized, and thus the surface of the platinum is kept clean, and the cell in condition to exert its maximum electro-motive power. In this combination we may substitute for the plate of platinum a plate of dense coke, such as forms in the interior of the gas retorts, which is very much cheaper, and enables us to construct large cells at a moderate cost. The use of gas coke was first suggested by Professor Bunsen of Heidelberg, and the cell so constructed generally bears his name. The Bunsen cell, such as is represented in Fig. 82, is exceedingly well adapted for use

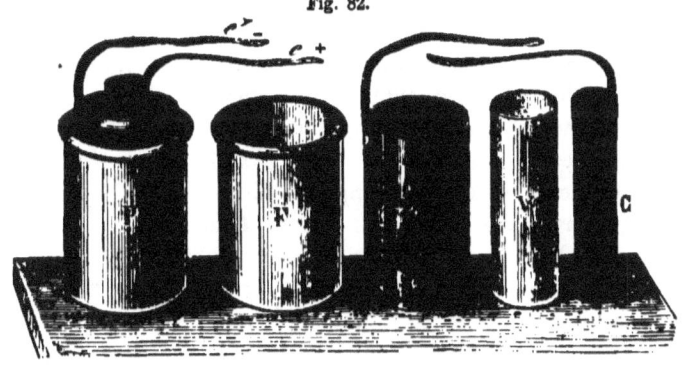

Fig. 82.

in the laboratory. These cells are usually made of nearly a

uniform size, the zinc cylinders being about 8 c. m. in diameter by 22 c. m. high, and they are frequently referred to as a rough standard of electrical power. They may be united so as to produce effects either of intensity or of quantity. The intensity effects are obtained in the manner already described (see Fig. 79), and the quantity effects are obtained with equal readiness; since by attaching the zinc of several cells to the same metallic conductor, and the corresponding coke plates to a similar conductor, we have the equivalent of one cell with large plates. Many other forms of battery, differing in more or less important details from those here described, and adapted to special applications of electricity, are used in the arts, and are fully described in the larger works on physics.

91. *Electrolysis.* — As has been already stated, the electrical current has the remarkable power of imparting to the unlike atoms of almost all compound bodies motion in opposite directions, like that in the battery cell itself, and this, too, at whatever point in the circuit they may be introduced. The galvanic battery thus becomes a most potent agent in producing chemical decompositions, and it is in consequence of this fact that the theory of the instrument fills such an important place in the philosophy of chemistry.

If we break the metallic conductor at any point of a closed circuit, the two ends, which in chemical experiments we usually arm with platinum plates,[1] are called poles. The end connected with the platinum or coke plate, from which the positive current is assumed to flow, is called the *positive pole*, and the end connected with the zinc plate, from which the negative current flows, is called the *negative pole*. Let us assume that Fig. 83 represents the two platinum poles dipping in a solution of hydrochloric acid in water, which thus becomes a part of the circuit. The moment the circuit is thus closed, the H and Cl atoms begin to travel in opposite directions, just as in the battery cell below. The hydrogen atoms move *with the positive current* towards the negative pole, and hydrogen gas is disengaged from the surface of

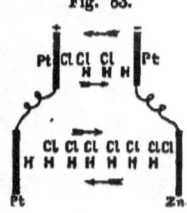

Fig. 83.

[1] We use platinum plates because this metal does not readily enter into combination with the ordinary chemical agents.

the negative plate, while the chlorine atoms move *with the negative current* towards the positive pole, and chlorine gas is evolved from the surface of the positive plate. Moreover, it will be noticed that each kind of atoms moves in the same direction on the closed circuit, that is, follows the course of the same current, both in the battery cell below and in the decomposing cell above; and wherever we break the circuit, and at as many places as we may break it, the same phenomena may be produced, provided only that our battery has sufficient power to overcome the resistance thus introduced.

If next we dip the poles in water, the atoms of the water will be set moving, as shown in Fig. 84; hydrogen gas escaping as before from the negative pole, and oxygen gas from the positive. We find, however, that pure water opposes a very great resistance to the motion of the current; and, unless the current has great intensity, the effects obtained are inconsiderable. But if we mix with the water a little sulphuric acid, the decomposition at once becomes very rapid; but then it is the atoms of the sulphuric acid, and not those of the water, which are set in motion. The molecule H_2SO_4 divides into H_2 and SO_4; the hydrogen atoms moving in the usual direction, and the atoms of SO_4 in the opposite direction. As soon, however, as the last are set free at the positive pole, they come in contact with water, which they immediately decompose, $2H_2O + 2SO_4 = 2H_2SO_4 + O_2$, and the oxygen gas thus generated escapes from the face of the platinum plate. Thus the result is the same as if water were directly decomposed, but the actual process is quite different.

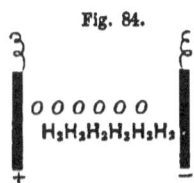
Fig. 84.

So also in many other cases of electrolysis, — as these decompositions by the electrical current are called, — the process is complicated by the reaction of the water, which is the usual medium employed in the experiments. Thus, if we interpose between the poles a solution of common salt, $NaCl$, the chlorine atoms move towards the positive pole, and chlorine gas is there evolved as in the first example. The sodium atoms move also, but in the opposite direction. As soon, however, as they are set free at the negative pole, they decompose the water present; hydrogen gas is formed, which escapes in bubbles from the platinum plate, while sodic hydrate (caustic soda) remains in solution,

$$2H_2O + 2Na = 2H, Na\text{-}O + H\text{-}H.$$

We add but one other example, which illustrates a very important application of these principles in the arts. We assume, in Fig. 85, that the positive pole is armed with a plate of copper, and that to the negative pole has been fastened a mould of some medallion we wish to copy, the surface of which, at least, is a good conductor. We assume further that both copper plate and mould are suspended in a solution of sulphate of copper, $Cu\text{=}SO_4$. In this case the atoms of the compound are set in motion as before. Those of copper accumulate on the surface of the mould; and at last the coating will attain such thickness that it can be removed, furnishing an exact copy of the original medallion. Meanwhile the atoms of SO_4 have found at the positive pole a mass of copper, with whose atoms they have combined; and thus fresh sulphate of copper has been formed, and the solution replenished. The process has in effect consisted in a transfer of metal from the copper plate to the medallion; and, by using appropriate solvents, silver and gold can be transferred and deposited in the same way.

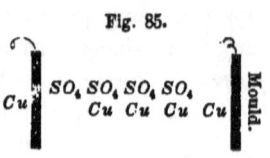

Fig. 85.

In all these processes of electrolysis, one remarkable fact has been observed, which has a very important bearing on the theory of the battery. If in any given circuit we introduce a number of decomposing cells, containing acidulated water, we find that in a given time exactly the same amount of gas is evolved in each; thus proving, what we have thus far assumed, that the moving power is absolutely the same at all points on the circuit. Moreover, the amount of gas which is evolved from such a decomposing cell in the unit of time is an accurate measure of the strength of the current actually flowing in any circuit, and this mode of measuring the quantity of an electrical current is constantly used.

We should infer from the facts already stated, and the principle has been confirmed by the most careful experiments, that the chemical changes which may take place at different points of the same closed circuit are always the exact equivalents of each other. If, for example, we have a series of Grove's cells, arranged as in Fig. 79, and interpose in the external circuit two decomposing cells, as in Figs. 84 and 85, we shall find (provided there is no local action) that the weight of zinc dissolved in each of the five Grove's cells is the exact *chemical equivalent*, (26) not only of the weight of hydrogen gas evolved from the first decomposing cell, but also of the weight of metallic copper deposited on the mould in the second. For every 63.4 grammes of copper deposited, 2 grammes of hydrogen are evolved, and 65.2 grammes of zinc are dissolved in *each* cell of the battery. If there is also local action in the cells, the chemical change thus induced is added to the normal effect of the battery-current.

The examples which have been given are sufficient to illustrate the remarkable power which the electric current possesses of setting in motion the atoms of compound bodies. Innumerable experiments have shown that, in reference to their relations to the current, the atoms, both simple and compound, may be divided into two great classes: first, those which travel on the line of the circuit in the direction of the positive current and follow in the lead of the hydrogen atoms; and, secondly, those which follow the lead of the chlorine atoms, and move in the opposite direction with the negative current. The first class of atoms, or radicals, we call *positive*; and the second class, *negative*.

The opposition in qualities of the chemical atoms, which the study of these electrical phenomena has revealed, is, in many cases at least, relative, and not absolute. For, while there are some atoms which always manifest the same character, there are others which appear in some associations positive, and in other associations negative. To such an extent is this true, that the electrical relations of the atoms are best shown by grouping the elements in series, which may be so arranged that each member of the series shall be electro-positive when in combination with those elements which follow it, and electro-negative when combined with those which precede it.

The simple mechanical theory of electrical currents which has been presented in this chapter, is adequate to explain the general order of the chemical phenomena with which we are

ELECTRICAL RELATIONS OF THE ATOMS. 169

more immediately concerned in this work. But there are other classes of electrical phenomena, of which this theory, at least in its present form, can give no account, and which have always been referred to the presence of an assumed electrical fluid, pervading all nature, and consisting of two oppositely polarized conditions of the same substance, — the vitreous and resinous, or positive and negative electricities, — which, when separated by chemical action, by friction, or in other ways, constantly tend to flow together through all those channels which we call electrical conductors. It is, however, the tendency of modern science to refer all physical changes to a simple mechanical cause, and although the phenomena of statical electricity are still best explained on the fluid hypothesis, we may hope that further study will show that they also may be reconciled with some dynamical theory. It is possible that the electrical fluid, which would seem to appear in these phenomena, is an "ethereal" atmosphere, surrounding the atoms, and that through this medium the electrical impulses are transmitted. (Compare 892.)

Questions and Problems.

In the following problems the values C, R or r and E of Ohm's formula are assumed to be measured in terms of the following units. First. The *unit of current* is that which would produce, by the electrolysis of water, 1 $\overline{c.m.}^{3}$ of hydrogen and oxygen gas (measured under standard conditions) in one minute. Secondly. The *unit of resistance* is that offered by a pure silver or copper wire 1 m. long and 1 m. m. diameter at 0°.[1] Lastly, the unit of *electromotive force* is that which transmits a *unit current* against a *unit resistance* in a *unit of time*.

1. What resistance does the current suffer in an iron wire 50 metres long and 5 m. m. diameter? Sp. R. of iron 7.

Ans. 14 units.

2. Assuming that the Sp. R. of copper is 1.3 and that of iron 7, what must be the diameter of an iron wire which will oppose no greater resistance to the current than a copper wire of 2 m. m. diameter?

Ans. 4.64 m. m.

[1] This unit is 0.02057 of the absolute unit recently adopted by the British Association.

170   ELECTRICAL RELATIONS OF THE ATOMS.

3. It is found by experiment that a wire of German silver, 7.201 m. long and 1.5 m. m. diameter, opposes the same resistance to the current as a wire of pure silver 10 m. long and ½ m. m. diameter. What is the Sp. R. of German silver.   Ans. 12.5.

4. It is required to make with 132.8 grammes of pure silver, a wire which will offer a resistance of 81 units. What must be its length and diameter? *Sp. Gr.* of silver = 10.57.

Solution. Representing by x the length in *metres*, and by y the diameter in *millimetres*, we deduce by [1] $y^2 x \frac{\pi}{4} 10.57 = 132.8$ and by the laws of conduction $\frac{x}{y^2} = 81$. Whence $x = 36$ m. and $y = \frac{2}{3}$ m. m.

5. What is the length and diameter of an iron wire weighing 97.38 grammes, which offers a resistance of 9,072 units? It is known that the Sp. Gr. of the iron = 7.75 and its Sp. R. = 7.

Ans. Length, 144 m.   Diameter, ⅓ m. m.

6. From a given wire there are four branches, of which the resistance is respectively 10, 20, 30, and 40. Required the total resistance when the current passes simultaneously through the four branches.

Solution. The resistance in the first branch may be represented by a normal silver wire 10 m. long and 1 m. m. diameter. If we call the area of a transverse section of this wire s, then the resistance in the other three branches will be represented by normal wires of the same length, but having on the cross sections the areas ½ s, ⅓ s and ¼ s respectively. If next we conceive of these wires as merged in one, having the common length 10 m. and an area on the section equal to $(1 + \frac{1}{2} + \frac{1}{3} + \frac{1}{4}) s$, it is evident that such a wire will represent the resistance required. Hence we easily deduce,

Ans. 4.8.

7. A closed circuit has two branches through which the current passes simultaneously. In one branch $r = 100$. What length of copper wire 5 m. m. diameter must be used for the other that the total $r = 50$?   Ans. 2,500 metres.

8. A conductor has two branches, one having $R = 756$, the other so adjusted that when the current passes at the same time through both, the *total* resistance equals 540. Required the length of a German silver wire ½ m. m. diameter and Sp. R. = 12.5, which, when inserted in the adjusted branch, will increase the *total* resistance to 630.

Solution. By principle of last problem we easily find that the resistance in the adjusted branch before insertion equals 1,890, and after insertion, 3,780. The difference between these values, 1,890, is the resistance due to the inserted wire. Hence its length must be 37.8 metres.

ELECTRICAL RELATIONS OF THE ATOMS.   171

9. We have a battery of six Daniells cells, in each of which $E = 475$, $R = 15$, and the external resistance against which the battery is to work, $r = 10$. The cells may be arranged, 1st, as six single elements; 2d, as three double elements;[1] 3d, as two three-fold elements; 4th, as one six-fold element. Required the current strength in each case.   Ans. 28.5, 43.8, 47.5 and 38.0 respectively.

10. We have a battery of twelve Grove cells, in each of which $E = 880$, and $R = 18$, to work against an external resistance of $r = 24$. Required the strength of current when the cells are arranged, 1st, as twelve single; 2d, as six two-fold; 3d, as four three-fold; 4th, as three four-fold; 5th, as two six-fold, and 6th, as one twelve-fold element.

Ans. 41.5, 63.8, 69.2, 66.4, 55.3, and 32.5 respectively.

11. With a single cell, where E and R have a constant value, what is the maximum strength of current, and under what conditions would it be obtained?

Ans. $\dfrac{E}{R}$, when the external resistance is nothing.

12. With n cells in each of which E and R have the same value, what is the maximum strength of current, and under what conditions would it be obtained?

Ans. $n\dfrac{E}{R}$, when the cells are arranged as one n-fold element, and work against no external resistance.

13. With n cells as above, working against a given external resistance r, how should they be arranged so as to obtain the maximum value of C?

Ans. So as to make the internal resistance equal to that of the external circuit.

Solution. If x represents the number of compound elements formed with the n cells when C in Ohm's formula is a maximum, we should evidently have under this condition x compound elements, each formed of $\dfrac{n}{x}$ cells. The electromotive force of such an arrangement would be xE. The internal resistance would be $xR \div \dfrac{n}{x} = \dfrac{x^2}{n}R$ (compare problems 8 and 9), and the strength of the maximum current required,

$$C = \dfrac{xE}{\dfrac{x^2}{n}R + r}$$

[1] By double elements is meant a group of two cells coupled for quantity [89] and equivalent to a large cell having plates of twice the size. Six double elements are six such groups arranged for intensity, and the other terms have a similar meaning.

The first differential coefficient of this function of x when C is a maximum must be equal to zero. Hence,

$$\frac{\left(\frac{x^2}{n}R+r\right)E - 2\frac{x^2}{n}RE}{\left(\frac{x^2}{n}R+r\right)^2} = 0$$

or
$$r = \frac{x^2}{n}R.$$

That is, the strength of the current is at its maximum when the internal equals the external resistance, as stated above. Those who are not familiar with the elementary principles of the differential calculus may satisfy themselves of the truth of this result by comparing the answers obtained to problems 8 and 9.

14. We have, in the first place, for a single cell of a given combination working against a feeble resistance, the value $C = \frac{E}{R+r}$; in the second place, for n cells of the same combination working against n times the resistance, the identical value $C = \frac{nE}{nR+nr}$. In "*strength*" the two currents are equal, but are they identical?

15. In a given cell $E = 475$; $R = 15$. The current passes through 30 metres pure copper wire 2 m. m. diameter. It is required to arrange 8 cells so that C may be the greatest possible.

Ans. They should be arranged as two four-fold elements.

16. We have a battery of four Bunsen cells ($E = 800, R = 4$ each), coupled as four single elements. The circuit is closed through 500 grammes of pure copper wire. Required the greatest strength of current, and the dimensions of the wire that this maximum may be obtained.

17. A simple Voltaic cell, whose electromotive force E is known, working against an unknown total resistance R' (both external and internal), produces a given effect upon a galvanometer. Another cell differently constructed, working against a total resistance R'', also unknown, produces the *same* effect upon the galvanometer. It is also observed that a measured length l of normal copper wire, inserted in the first circuit, produces on the galvanometer the same difference of effect as a length l' inserted in the second circuit. Required the electromotive force E' of the second cell.

Solution. We easily deduce from Ohm's formula the two equations $\frac{E}{R'} = \frac{E'}{R''}$ and $\frac{E}{R'+l} = \frac{E'}{R''+l'}$, whence we obtain, —

Ans. $E' = E\frac{l'}{l}.$

ELECTRICAL RELATIONS OF THE ATOMS. 173

18. In order to determine the electromotive force of a Bunsen's cell, it was compared, as in last problem, with a Daniell's cell whose electromotive force was known to be 470. After adjusting the external resistances so that both produced the same effect upon the galvanometer, it was found that the insertion of 5.6 m. of copper wire into the first circuit caused the same change in the instrument as the insertion of 3.29 metres of the same wire in the circuit of the Daniells cell. What was the electromotive force sought?
Ans. 800.

19. A battery of 40 Bunsen's cells remains closed for an hour, and during that time furnishes a current whose strength $C = 30$. How much zinc will be consumed in this time, assuming that there is no local action?

Solution. Such a current would produce, by the electrolysis of water, 30 $\overline{c. m.}^3$ of gas in one minute, or 1.8 litres in one hour. Of this gas 1.2 litres or 1.2 criths would be hydrogen. The chemical equivalent of zinc being 32.6, the amount of zinc dissolved in each cell must be $1.2 \times 32.6 = 39.12$ criths, and in the forty cells 1564.8 criths, equal to 140 grammes, the answer required.

20. In an electrotype apparatus, Fig. 85, 16.36 grammes of copper were deposited on the negative mould in 24 hours. What was the strength of current? Ans. 6 units.

21. In an electrotype apparatus the electromotive force of the single cell employed is 420, and the internal resistance 5. The external resistance, including decomposing cell, is 0.25. How much copper will be deposited on the negative mould in one hour, and how much zinc will be dissolved in the battery during the same time? Ans. 9.088 grammes copper and 9.346 grammes of zinc.

22. Thirty-two Grove cells ($E = 830$, $R = 20$ each) are connected as 4 eight-fold compound elements and the current employed to work an electro-silvering apparatus, in which the total resistance external to the battery was equivalent to 10. Required the number of grammes of silver deposited each hour, and the number of grammes of zinc dissolved during the same time in the battery.
Ans. 64.24 grammes of silver and 77.56 grammes of zinc.

23. Assuming that the external resistance cannot be changed, would the same number of cells of the battery described in last problem be so arranged as to deposit more silver in the same time?
Ans. They could not.

Could they be so arranged as to deposit the same amount of silver with less expense of zinc? What would be the most economical arrangement, and under these conditions how much silver would be deposited in one hour and how much zinc dissolved?

Answer to last question, 30.25 grammes silver, and 9.13 grammes of zinc.

CHAPTER XVI.

RELATIONS OF THE ATOMS TO LIGHT.

92. *Light a Mode of Atomic Motion.* — It has already been intimated (§ 53, note), that the phenomena of vision are the effects of an atomic motion transmitted from some luminous body to the eye through continuous lines of material particles, and such lines we call rays of light. This motion may originate with the atoms of *various* substances; but in order to explain its transmission, we are obliged to assume the existence of a medium filling all space, of extreme tenuity, and yet having an elasticity sufficiently great to transmit the luminous pulsations with the incredible velocity of 186,000 miles in a second of time. This medium we call the ether, but of its existence we have no definite knowledge except that obtained through the phenomena of light themselves, and these require assumptions in regard to the constitution of the ethereal medium which are not realized even approximately in the ordinary forms of matter; for while the assumed medium must be vastly less dense than hydrogen, its elasticity must surpass that of steel.

According to the undulatory theory, motion is transmitted from particle to particle along the line of each luminous wave very much in the same way that it passes along the line of ivory balls in the well-known experiment of mechanics. The ethereal atoms are thus thrown into waves, and the order of the phenomena is similar to that with which all are familiar in the grosser forms of wave motion. But in this connection we have no occasion to dwell on the mechanical conditions attending the transmission of the motion. The motion itself may be best conceived as an oscillation of each ether particle in a plane perpendicular to the direction of the ray, not necessarily, however, in a straight line; for the orbit of the oscillating molecule may be either a straight line, an ellipse, or a circle, as the case may be. Such oscillations may evidently differ both as regards their amplitude and their duration, and on these fundamental elements depend two important differences in the effect of the motion on the organs of vision, viz. intensity and quality, or brilliancy and color.

If our theory is correct, it is obvious that the intensity of the luminous impression must depend upon the force of the atomic blows which are transmitted to the optic nerves, and it is also evident that this force must be proportional to the square of the velocity of the oscillating atoms, or what amounts to the same thing, to the square of the amplitude of the oscillation; assuming, of course, that the oscillations are isochronous.

The connection of color with the time of oscillation is not so obvious, and why it is that the waves of ether beating with greater or less rapidity on the retina should produce such sensations as those of violet, blue, yellow, or red, the physiologist is wholly unable to explain. We have, however, an analogous phenomenon in sound, for musical notes are simply the effects of waves of air beating in a similar way on the auditory nerves; and, as is well known, the greater the frequency of the beats, or, in other words, the more rapid the oscillations of the aerial molecules, the higher is the pitch of the note. Red color corresponds to low, and violet to high notes of music, and, the gradations of color between these extremes, passing through various shades of orange, yellow, green, blue, and indigo, correspond to the well-known gradations of musical pitch.

From well-established data we are able to calculate the rapidity of the oscillations which produce the different sensations of color, and also to estimate the corresponding lengths of the ether waves, and the following table contains the results. It must be understood, however, that these numbers merely correspond to a few shades of color definitely marked on the solar spectrum by certain dark lines hereafter to be mentioned; and that equally definite values may be assigned to the infinite number of intermediate shades which intervene between these arbitrary subdivisions of the chromatic scale.

176 RELATIONS OF THE ATOMS TO LIGHT.

Color.	Number of waves or oscillations in one second.		Length of waves in fractions of a millimetre.
Red ·	477 million	million.	650 millionths.
Orange	506	"	609 "
Yellow	535	"	576 "
Green	577	"	536 "
Blue	622	"	498 "
Indigo	658	"	470 "
Violet	699	"	442 "

93. *Natural Colors.* — It follows, as a necessary consequence of the fundamental laws of mechanics, that an oscillating molecule can only transmit to its neighbor motion which is isochronous with its own. Hence a single ray of light can only produce a definite effect of color, and this quality of the ray will be preserved however far the motion may travel. A beam of light is simply a bundle of rays, and if the motion is isochronous in all its parts, that is, if the beam consists only of rays of one shade of color, such a beam will produce the simplest chromatic sensation possible, — that of a pure color. If, however, the beam contains rays of different colors, we shall have a more complex effect, and the infinite variety of natural tints are thus produced. When, lastly, the beam contains rays of all the colors mingled in due proportion, we receive an impression in which no single color predominates, and this we call white light.

The colors of natural objects, whether inherent or imparted by various dyes, are simply effects upon the retina produced by the beam after it has been reflected from the surface or transmitted through the mass of the body, and the peculiar chromatic effects are due to the unequal proportions in which the different colored rays are thus absorbed. The color reflected, and that absorbed or transmitted, are always complementary to each other, and if mingled they would reproduce white. It is obvious, moreover, that no beam of light, however modified by reflection or transmission, could produce the sensation of a given color, if it did not contain from the first the corresponding colored rays. Hence it is that the colors of objects only appear naturally by daylight, and when illuminated by a monochromatic light, all colors blend in that of this one pure tint.

94. *Chromatic Spectra and Spectroscopes.* — When a beam of light is passed through a glass prism placed as shown in Fig.

RELATIONS OF THE ATOMS TO LIGHT. 177

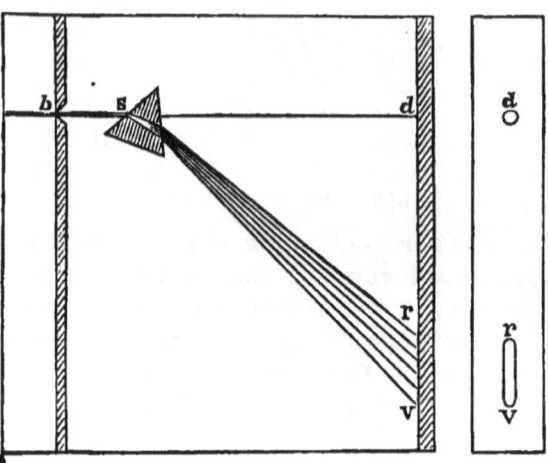

Fig. 86.

86, it is not only *refracted*, that is, bent from its original rectilinear course, but the colored rays of which the beam consists, being bent unequally, are separated to a greater or less extent, and falling on a screen produce an elongated image colored with a succession of blending tints, which we call the spectrum. The red rays, which are bent the least, are said to be the *least refrangible*, while the violet rays are the *most refrangible*, and intermediate between these we have, in the order of refrangibility, the various tints of orange, yellow, green, blue, and indigo. Thus a prism gives an easy means of analyzing a beam of light, and of discovering the character of the rays by which a given chromatic effect is produced. Such observations are very greatly facilitated by a class of instruments called spectroscopes, and Figs. 87 and 90 will illustrate the principles of their construction.

In the very powerful instrument first represented, the beam of light is passed in succession through nine prisms (each having an angle of 45°), and the extreme rays are thus widely separated, while the beam itself is bent around nearly a whole circumference. The only other essential parts of the instrument are the collimator A and the telescope B. The light first enters the collimator through a narrow slit, and having passed through the prisms is received by the telescope. The telescope is adjusted as it would be for viewing distant objects,

and a lens at the end of the collimator serves to render the
rays diverging from the slit parallel, so that when the two

Fig. 87.

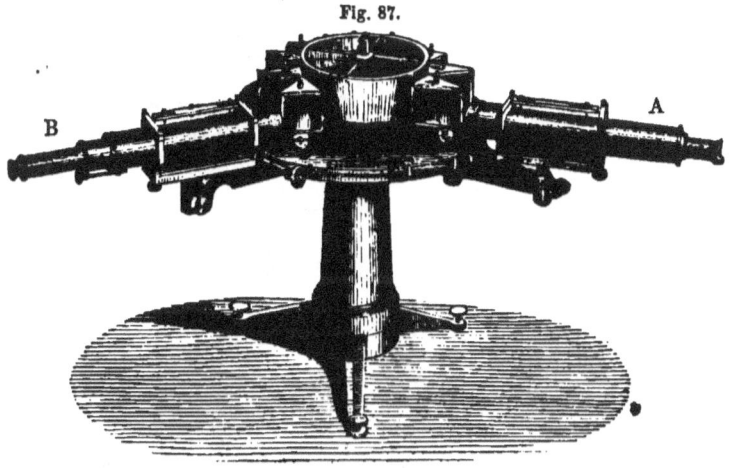

tubes are in line, one sees through the telescope a magnified image of the slit, just as if the slit were at a great

Fig. 88.

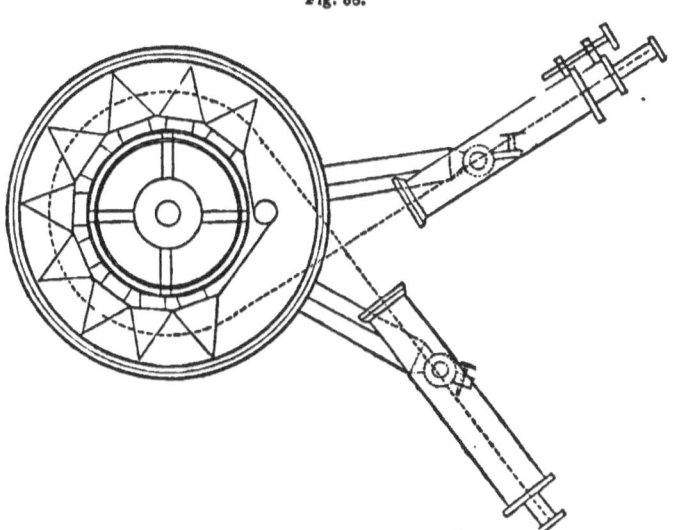

distance. In like manner when the telescopes are placed as
in Fig. 88, and when the light before reaching the telescope

RELATIONS OF THE ATOMS TO LIGHT. 179

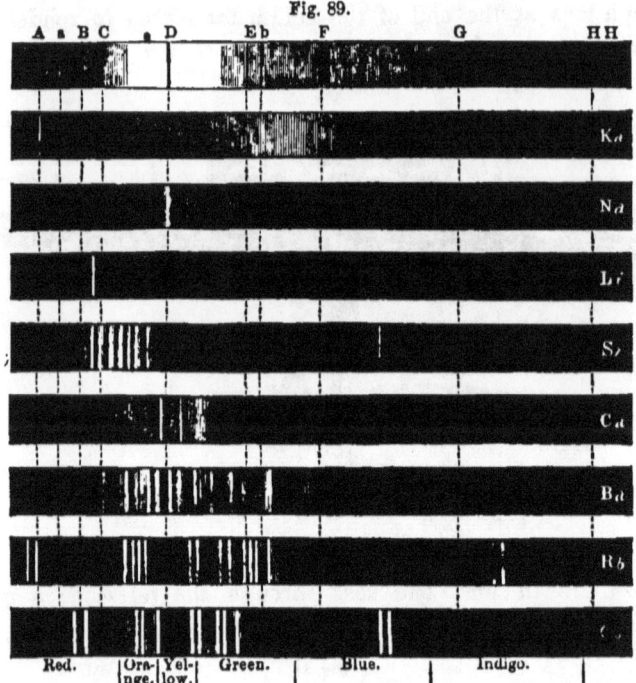

Fig. 89.

passes through the whole series of prisms, we still see a single definite image whenever the slit is illuminated by a pure monochromatic light. Moreover, this image has a definite position in the field of view, which, when the instrument is similarly adjusted, depends solely on the refrangibility of the light.

Thus, if we place in front of the slit a sodium flame, which emits a pure yellow light, we see a single yellow image of this longitudinal opening, as in Fig. 89, Na. If we use a lithium flame, we see a similar image,[1] but colored red, and at some distance from the first, to the left, if the parts of our instrument are disposed as in Fig. 88. If we use a thalium flame, we in like manner see a single image, but colored green, and falling considerably to the right of both of the other two. If now we illuminate the slit by the three flames simultaneously, we see all three images at once in the same relative position as before. So also if we examine the

[1] The second image shown in Fig. 89, Li is not ordinarily seen.

flame of a metal, which emits rays of several definite degrees of refrangibility, we see an equal number of definite images of the slit. If, next, we illuminate the slit with sunlight, which contains rays of all degrees of refrangibility, we see an infinite number of images of the slit spread out along the field of view, and these, overlapping each other, form that continuous band of blending colors which we call the solar spectrum. If, lastly, we examine with our instrument the light reflected from a colored surface, or transmitted through a colored medium, we also see a band of blending colors, but at the same time we observe that certain portions of the normal solar spectrum are either wholly wanting or greatly obscured.

With a spectroscope of many prisms like the one represented by Fig. 87, we can only see a small portion of the spectrum at once. By moving the telescope, which, fastened to a metallic arm, revolves around the axis of the instrument, different portions of the spectrum may be brought into the field of view; while a vernier, attached to the same arm and moving over a graduated arc, enables us to fix the position of the spectrum lines, as the images of the slit are usually called. The other mechanical details shown in the figure are required in order to adjust the various parts of the instrument, and especially in order to bring the prisms to what is termed the angle of minimum deviation. But an instrument of this magnitude and power is not required for the ordinary applications of the spectroscope in chemistry. For this purpose a small instrument consisting of a collimator, a single prism, and a telescope, all in a fixed position, are amply sufficient. In the field of such a spectroscope the whole spectrum may be seen at once; and the position of the spectrum lines is very easily determined by means of a photographic scale placed at one side, and seen by light reflected into the telescope from the face of the prism.

The various parts of the instrument, as arranged for observation, are shown in Fig. 90. A is the collimator, P the prism, and B the telescope. The tube C carries the photographic scale, and has at the end nearest to the prism a lens of such focal length that the image both of the slit and the scale may be seen through the telescope at the same time, the one appearing projected upon the other. The screw e serves to adjust the width of the slit. Moreover, one half of the

Fig. 90.

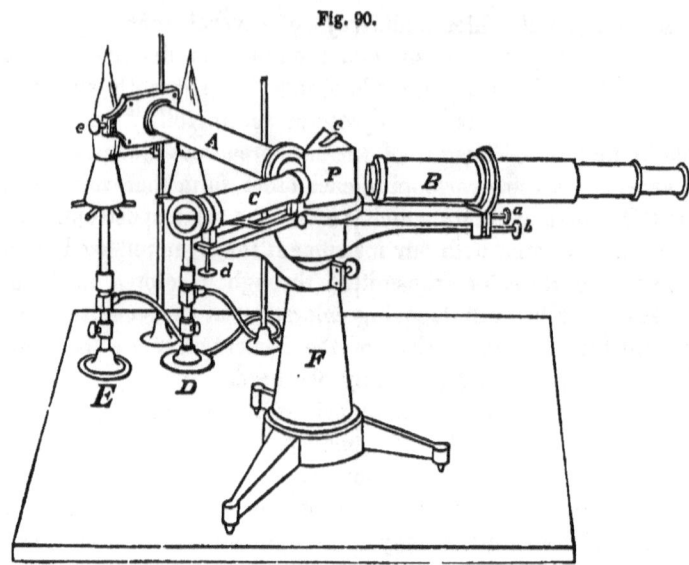

length of the slit is covered by a small glass prism so arranged that it reflects into the collimator tube the rays from a lamp placed on one side. Thus the two halves of the slit may be illuminated independently by light from different sources, and the two spectra, which are then seen superimposed upon each other (see Fig. 91), exactly compared. The various screws, which appear in Fig. 90, are used for adjusting the different parts of the instrument.

95. *Spectrum Analysis.* — The atoms of the different chemical elements, when rendered luminous under certain definite conditions, always emit light whose color is more or less characteristic, and which, when analyzed with the spectroscope, exhibit spectra similar to those which are represented in Fig. 89, so far as is possible without the aid of color. Sometimes we see only a single line in a definite position, as in the case of Na, Li, and Th, already referred to. At other times there are several such lines; and, still more frequently, to these lines (or definite images of the slit) there are superadded more or less extended portions of a continuous spectrum. Moreover, not only is the general aspect of each spectrum exceedingly characteristic, but also the occurrence of its peculiar lines is, so far as we know, an absolute proof of the

presence of a given element, and these lines may be readily recognized by their position, even when the character of the spectrum is otherwise obscure. It is evident then that we have here a principle which admits of most important applications in chemical analysis, and it only remains to consider under what conditions the elementary atoms emit their characteristic light.

First. All bodies when intensely heated are rendered luminous, and, other things being equal, the higher the temperature the more intense is the light. The brilliancy of the light emitted at the same temperature by different bodies varies very greatly, the densest bodies being, as a general rule, the most intensely luminous.

Secondly. — Solid and liquid bodies, if opaque, emit when ignited white light, or at least light which shows with the spectroscope a continuous spectrum more or less extended. At a red heat the light from such bodies consists chiefly of red rays, but as the temperature rises first to a white and then to a blue heat, the more refrangible rays become more abundant and finally predominate.

Thirdly. — The elementary substances give out their peculiar and characteristic light only in the state of gas or vapor. Hence, when we examine with a spectroscope a source of light, we may infer that a continuous spectrum indicates the presence of solid or liquid bodies, while a discontinuous spectrum, with definite lines or images of the slit, indicates the presence of gases and vapors; and in the last case we can, as has been seen, infer from the position of the lines the nature of the luminous atoms. It would seem, however, from recent investigations, that under certain conditions even a gas may show a continuous spectrum, and there are other seeming exceptions which admonish us that the general principles just stated should be applied with caution.

Fourthly. — At the very high temperatures at which alone gases or vapors become luminous, compound bodies, as a rule, appear to be decomposed, and the elementary atoms disassociated. Hence the observations with the spectroscope have been almost entirely confined to the spectra of the elementary substances, and our knowledge of the spectra of compound substances is exceedingly limited. In some few cases where the spectrum of a compound has been obtained, it has been noticed that, as the temperature rises, this spectrum is suddenly resolved into the separate spectra of the elements of which the compound consists.

Fifthly. — At a high temperature the metallic atoms of a compound body are far more luminous than those of the other elementary atoms with which they are associated. Hence, when the vapor of a metallic compound is rendered luminous, the light emitted is so exclusively that of the metallic atoms, disassociated by the heat, that when examined with the spectroscope the spectrum of the metal is alone seen; and this is the probable explanation of the fact that the salts of the same metal, when treated as will be described in the next paragraph, all show, as a general rule, the same spectrum as the metal itself.

Lastly. — The substance, on which we wish to experiment, may be rendered luminous in several ways. If the substance is a volatile metallic salt, the simplest method is to expose a bead of the substance (supported on a loop of platinum wire) to the flame of a Bunsen's burner (Fig. 90), which by itself burns with a nearly non-luminous flame. The flame soon becomes filled with the disassociated atoms of the metal and shines with their peculiar light.

In order to study the spectra of the less volatile metals like aluminum, iron, or nickel, we use two needles of the metal, and pass between the points, when about one fourth of an inch apart, the electric discharges of a powerful Ruhmkorff coil, condensed by a large Leyden jar. The metal is volatilized by the heat of the electric current, and the space between the points becomes filled with the intensely ignited vapor, which then shines with its characteristic light."[1]

In a similar way we can experiment on the permanent gases and lighter vapors, enclosing them in a glass tube with platinum electrodes, and before sealing the tube reducing the tension with an air pump, when the discharge will pass through a length of several inches of the attenuated gas. The light then emitted comes from the atoms or molecules of the gas, and where the electric current is condensed as in the capillary por-

[1] An electric spark is in every case merely a line of material particles rendered luminous by the current.

tion of the tubes constructed for this purpose, the light is sufficiently intense to be analyzed with the spectroscope.

The three different modes of experimenting just described do not by any means always give the same spectrum when applied to the same chemical element. It constantly happens that as the temperature rises new lines appear, which are usually those corresponding to the more refrangible rays, and at the very high temperatures generated by the electric discharge many of the spectra change their whole aspect. The ill-defined broad bands or luminous spaces which are so conspicuous at a low temperature (see Fig. 89), disappear, and are replaced by a greater or less number of definite spectrum lines. Generally, however, the characteristic lines which mark the element at the lower temperature are seen also at the higher; but sometimes there is a sudden and complete change of the whole spectrum. The cause of these differences is not understood, but it has been thought by some investigators that the normal spectra of the elementary atoms consist of bright bands alone, and that the more or less continuous spectra, which are also seen at the lower temperatures, are to be referred to the imperfect disassociation of the atoms, whose mutual attractions or partial combinations produce a state of aggregation approaching the condition which determines the continuous spectra of liquid or solid bodies.

96. *Delicacy of the Method.* — Having now stated the general principles of spectrum analysis, and the conditions under which these principles may be applied, it need only be added that the method is one of extreme delicacy. It enables us to detect wonderfully minute quantities of many of the metallic elements, and has already led to the discovery of four elements of this class which had eluded all methods of investigation previously employed. The names of these elements, Rubidium, Caesium, Thallium and Indium, all refer to the color of their most characteristic spectrum bands.[1]

97. *Solar and Stellar Chemistry.* — When a beam of sunlight is examined with a powerful spectroscope, the solar spectrum is seen to be crossed by an almost countless number of *dark lines* distributed with no apparent regularity, and dif-

---

[1] The different bands of the same element are usually distinguished by Greek letters, following the order of relative brilliancy.

fering very greatly in relative strength or intensity. These lines were first accurately described by the German optician Fraunhofer, and have since been known as the Fraunhofer lines. A few of the most prominent of these lines are shown in Fig. 89, with the letters of the alphabet by which they are designated. These lines, like the bright lines of the elements, correspond in every case to a definite degree of refrangibility, and therefore have a fixed position on the scale of the spectroscope. Moreover, what is very remarkable, the bright and the dark lines have in several cases absolutely the same position.

It is easy to construct the spectroscope so that the two halves of the slit may be illuminated from different sources. If then we admit a beam of sunlight through one half, and the light of a sodium flame through the other half, we shall have the two spectra super-imposed in the same field, as in Fig. 91,

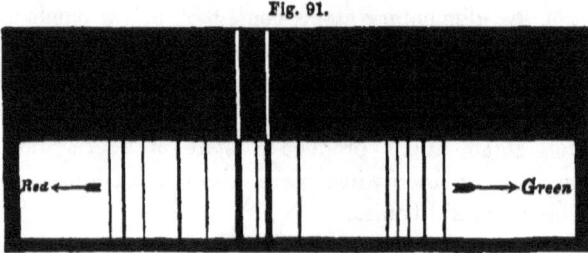

Fig. 91.

and it will be seen that the two parts of the sodium band, which appears as a double line under a high power, coincide absolutely in position with the double dark line D in the solar spectrum. But a still more striking coincidence has been observed in the case of iron, for the eighty well-marked bright lines in the spectrum of this metal correspond absolutely both in position and in strength with eighty of the dark lines of the solar spectrum. Now, the chances that such coincidences are the result of accident, are not one in one billion billion; and we are therefore compelled to believe that the two phenomena must be connected. A simple experiment shows what the relation probably is.

If we place before the spectroscope a sodium flame, we see, of course, the familiar double line. If now we place behind the sodium flame a candle flame, so that the candle also shines into the slit, but only through the sodium flame, we shall see the same bright lines projected upon the continuous spectrum of the candle. If, however, we put in place of the candle an electric light, we shall find that while the continuous spectrum is now far more brilliant than before, the sodium lines appear black. The explanation of this singular phenomenon is to be found in a principle, now well established both theoretically and experimentally, that a mass of luminous vapor, while otherwise transparent, powerfully absorbs rays of the same refrangibility which it emits itself. Hence, in our experiment, the very small portion of the spectrum covered by the sodium line is illuminated by the sodium flame alone, while all the rest of the spectrum is illuminated from the source behind, and the effect is merely one of contrast, the sodium lines appearing light or dark according as they are brighter or darker than the contiguous portions of the spectrum.

In a similar way the bright lines of a few other elements have been inverted, and these experiments would lead us to infer that the Fraunhofer lines themselves are formed by a brilliant photosphere shining through a mass of less luminous gas. In other words, it would appear that the sun's luminous orb is surrounded by an immense atmosphere which intercepts a portion of his rays, and that we see as dark lines what would probably appear as bright bands, could we examine the light from the atmosphere alone.

If then our generalization is safe, the dark and the bright lines are the same phenomena seen under a different aspect, and the one as well as the other may be used to identify the different chemical elements. Hence, then, there must be both iron and sodium in the sun's atmosphere, and for the same reason we conclude that our luminary must contain Calcium, Magnesium, Nickel, Chromium, Barium, Copper, and Zinc, while there is equally good evidence that Gold, Silver, Mercury, Aluminum, Cadmium, Tin, Lead, Antimony, Arsenic, Strontium, and Lithium are not present, at least in large quantities. It is true, however, that the elements thus recognized in the sun only account for a very insignificant portion of the dark lines, and it is difficult to reconcile this fact with our actual knowledge and present theories. Since the meteorites have brought to us no new elements, their evidence, as far as it goes, would not lead us to expect to find in the sun's atmosphere such a vast number of unknown elements as the dark lines would indicate; and this obvious explanation of their countless number cannot therefore be regarded as probable. It has been observed, however, in the few cases which have been investigated, that the spectrum of a compound body is far more complex than the different spectra of its elements combined; and it is possible that the complexity we see in the solar spectrum may arise from the partial combination or mutual interference of elements now known, in the outer and colder portions of that immense atmosphere which is supposed to extend many thousand miles beyond the luminous surface of the sun.

If next we turn the spectroscope on some of the brighter fixed stars, we shall see continuous spectra like the solar spectrum, of greater or less extent, and covered by dark lines. A careful comparison of these lines would seem to indicate that the stars differ very greatly from each other, although in general they are bodies similar to our sun; and if our theory is correct, we have been able to detect the presence of sodium, magnesium, hydrogen, calcium, iron, bismuth, tellurium, antimony, and mercury in Aldebaran, and other elements in other stars.

The most remarkable result of stellar chemistry remains yet to be noticed. On examining the nebulæ with the spectroscope, it has been found that while some of them show a continuous spectrum, there are a number of these remarkable bodies which exhibit the phenomena of bright lines. This would lead us to the conclusion that the last are really, as the nebular theory assumes, masses of incandescent gas, while the first are not true nebulæ, but simply clusters of very distant stars. An examination of the comets has confirmed the previous conclusion that they also are mere masses of gas, but, singularly enough, the light from the coma of one of those bodies gave a continuous spectrum, due probably to reflected sunlight.

98. *Absorption Spectra.* — When a luminous flame is viewed with a spectroscope through a solution of any salt of the metal *Erbium*, the otherwise continuous spectrum of the flame

is seen to be interrupted by several broad bands, which have a definite position, and are a valuable means of recognizing the presence of this very rare element. This absorption spectrum, as it is called, is simply the reverse, the "negative" of the luminous spectrum of the same element.

In like manner the salts of *Didymium* give an equally characteristic, although very different, absorption spectrum, which is in fact the only sure test we possess for this remarkable elementary substance; and as the bands may under some conditions be seen with reflected, as well as with transmitted light, we may apply the test even to opaque solids. Also, the same absorption bands are obtained either when the light is transmitted through a *liquid* solution, or through a *solid* crystal of *any* salt of the metal; and, moreover, the incandescent vapor of the metal shows bright bands corresponding to the dark bands in position. These facts would seem to show that the characteristic spectrum bands of an element may be, at least to some extent, independent both of the state of aggregation, and of the condition of combination of the elementary atoms.

Many substances besides the compounds of the elements just noticed, give characteristic absorption spectra which have been found to be useful chemical tests, especially in the case of blood, and certain other bodies of organic origin. The most remarkable phenomena of this class are the absorption spectra which are seen when a luminous flame is viewed with a spectroscope through various colored vapors, such as those of nitric per-oxide, bromine, and iodine. The dark bands are then very numerous, and in some cases may be resolved into well-defined lines. Indeed, the absorption bands are a class of phenomena closely allied to the Fraunhofer lines, many of which are known to result from the absorption by the earth's atmosphere of solar rays of certain degrees of refrangibility; and all these facts, with many others, prove that gases and vapors may exert their peculiar power of elective absorption at the ordinary temperature, as well as when incandescent. As a general rule, however, the absorption bands are not, like the bright lines of the metallic spectra or their representatives among the dark lines of the solar spectrum, definite images of the slit, but they are darker portions of the spectrum more or less regularly shaded, and correspond to the broad bands or luminous spaces in the spectra of the metallic vapors when not intensely heated. In each case the effect results from the blending of a greater or less number of images of the slit, differing in relative position and intensity.

99. *Theory of Exchanges.* — The facts of the two last sections are all illustrations of a general principle already referred to in connection with the reversal of the sodium spectrum. This principle is known as the "Theory of Exchanges," and has been stated as follows: "The relation between the power of emission, and power of absorption *for each kind of rays* (light or heat) is the same for all bodies at the same temperature." .... "Let R denote the intensity of radiation of a particle for a given description of light at a given temperature, and let A denote the proportion of rays of this description incident on the particle which it absorbs; then $R \div A$ has the same value for all bodies at the same temperature, — that is to say, this quotient is a function of the temperature only."

The law of exchanges finds its widest application in the phenomena of radiant heat, and so far as experiments have been made, it appears to be true in its greatest generality. In applying it to explain the reversal of the spectra of colored flames, we have only to deal with a single body in its relations to rays of different qualities. If the principle is true, the absorbing power of such a body at a given temperature must bear a fixed ratio to its power of emission for each kind of ray. If, for example, it has a great power of emitting certain rays of red light, it has a proportionally great power of absorbing the same rays. If, again, it has a feeble power of emitting violet rays of definite quality, its power of absorbing such rays is proportionally feeble, and bears the same ratio to the power of emission as before; and, lastly, it has no power of absorption over such rays as it does not itself emit. Moreover, it would follow that, although the relation of the absorbing to the radiating power might vary very greatly, so that, as the temperature falls, the last may become inconsiderable as compared with the first, or even vanish, no essential change in the character of the elective absorption would be thus induced. Hence, we should expect that bodies would absorb when cold rays of the same quality which they emit when hot, and also that opaque solids when heated would emit white light. We have seen that the general order of the phenomena is that which the law of exchanges would predict, and here, for the present, our knowledge stops. We have as yet been able to form no satisfactory theory in regard to the relations of the molecular structure of bodies to the medium through which the waves of light or heat are transmitted. It is, however, worthy of notice that Euler, one of the earliest and ablest investigators of undulatory motion, predicted the discovery of the law of exchanges, in assuming as a fundamental principle of the undulatory theory that a body can only absorb oscillations isochronous with these of which it is itself susceptible.

100. *General Conclusions.* — The facts that have been stated in this chapter are sufficient to show, that, although yet in its infancy, spectrum analysis promises to be one of the most powerful instruments of investigation ever applied in physical science. It seems to be the key which will in time open to our view the molecular structure of matter; and even now the results actually obtained suggest speculations in regard to the ultimate constitution of matter, of the most interesting character. The several monochromatic rays which the atoms of the elements emit, must receive their peculiar character from some motion in the atoms themselves which is isochronous with the motion they impart. Is it not then in this motion that the *individuality* of the element resides, and may not all matter be alike in its ultimate essence? Such speculations, however wild, are not wholly unprofitable, if only they stimulate investigation and thus lead to further discoveries.

CHAPTER XVII.

CHEMICAL CLASSIFICATION.

101. *General Principles*. — The glimpses that we have been able to gain of the order in the constitution of matter give us grounds for believing that there is a unity of plan pervading the whole scheme, and encourage a confident expectation that hereafter, when our knowledge becomes more complete, chemists may attain to at least such a partial conception of this plan as will enable them to classify their compounds under some natural system; and in imagination we may even look forward to the time when science will be able to express all the possibilities of this scheme with a few general formulæ, which will enable the chemist to predict with absolute certainty the qualities and relations of any given combination of materials or conditions. But although to a very slight extent the idea has been realized for a small class of the compounds of carbon, yet as a whole this grand conception is as yet but a dream. The more advanced student will find that in limited portions of some few fields of investigation a fragmentary classification is possible, as in mineralogy; but, when he attempts to comprehend the whole domain, he becomes painfully aware of the immense deficiencies of his knowledge; he is confused by the numerous chains of relationship, which he follows, with no result, to sudden breaks, and soon becomes convinced that all such efforts must be fruitless until more of the missing links are supplied.

The best that can now be done in an elementary treatise on chemistry is to group together the elements, or, rather, the elementary atoms, in such families as will best show their natural affinities; and then to study, under the head of each element, the more important and characteristic of its compounds. However little value such a classification may have in its scientific aspect, it will bring together, to a greater or less extent, the allied facts of the science, and thus will help the mind to retain them in the memory.

In classifying the elementary atoms, the three most important characters to be observed are the *Prevailing Quantivalence,* the *Electrical Affinities,* and the *Crystalline Relations.* The first of these characters serves more particularly to classify the elements in groups, the second to determine their position in the groups, and the last to control the indications of the other two.

The crystalline relations of the atoms can only be determined by comparing the crystalline forms of allied compounds, and involve the principles of isomorphism already discussed. Moreover, in order to reach the most satisfactory scheme of classification, we must take into consideration other properties of these compounds besides the crystalline form; which, although they may not be so precisely formulated, are frequently important aids in forming correct opinions as to the relations of the atoms. It will also be evident, from what has previously been stated, that more trustworthy inferences as to these relations may frequently be drawn from the crystalline form and properties of allied compounds than from those of the elementary substances themselves; for, in addition to the fact that so many of these substances crystallize in the isometric system, whose dimensions admit of no variation, it is also true that, in our ignorance of the molecular constitution of most of them, we often have more certainty, in the case of compounds, that our comparisons are made under identical molecular conditions.

102. *Metallic and Non-Metallic Elements.* — In all works on chemistry since the time of Lavoisier, the elementary substances have been divided into two great classes, — the *metals* and the *non-metals;* and the distinction is undoubtedly fundamental, although too much importance has been frequently attached to the accident of a brilliant lustre. The characteristic qualities of a metal, with which every one is more or less familiar, are the so-called *metallic lustre,* that peculiar adaptability of molecular structure known as *malleability* or *ductility,* and the *power of conducting electricity or heat.* These qualities are found united and in their perfection only in the true metals, although one or even two of them are well developed in several elementary substances which, on account of their chemical qualities, are now almost invariably classed with the non-metals, — as, for example, in selenium, tellurium, arsenic,

antimony, boron, and silicon. Besides the properties above named, many persons also associate with the idea of a metal a high specific gravity; but this property, though common to most of the useful metals, is by no means universal; and, among the metals with which the chemist is familiar, we find the lightest, as well as the heaviest, of solids. The non-metallic elements, as the name denotes, are distinguished by the absence of metallic qualities; but the one class merges into the other.

The presence or absence of metallic qualities in the elementary substances is for some unknown reason intimately associated with the electrical relations of their atoms, — those of the metals being electro-positive, while those of the non-metals are electro-negative, with reference, in each case, to the atoms of the opposite class. In the classification given in Table II. we have associated together in the same family both the metals and the non-metals having the same quantivalence, believing that such an arrangement not only best exhibits the relations of the atoms, but also that in a course of elementary instruction it presents the facts of chemistry in the most logical order.

103. *Scheme of Classification.* — The classification of the elementary atoms which has been adopted in this book is shown in Table II.

In the first place the atoms are divided into two large families, the Perissads and the Artiads (27).

Secondly, these families are subdivided into groups (separated by bars in the table) of closely allied elements. The atoms of any one of these groups are isomorphous; and they are arranged in the order of their weights, which is found to correspond also, in almost every case, to their electrical relations. Each group forms a very limited chemical series; and not only the weights and the electrical relations of the atoms, but also many of the physical qualities of the elementary substances, vary regularly as we pass from one end of the series to the other. The order of the variation, however, is not always the same; for while in some cases the lightest atoms of a series are the most electro-negative, in other cases they are the most electro-positive.

Thirdly, in arranging the groups of allied atoms we have followed the prevailing quantivalence of the group, and those groups whose elementary atoms exhibit in general the lowest quantivalence are, as a rule, placed first in order; but with our present limited knowledge there must be some uncertainty in regard to the details of such an arrangement, and the principle has sometimes been violated so as to bring together those groups of atoms which are most allied in their chemical relations.

The remarks already made in regard to the general scheme of chemical classification apply with almost equal force to the partial system here attempted. The very attempt makes evident the fragmentary character of our knowledge, even in regard to the exceedingly limited portion of the subject with which we are dealing. The idea of classification by series was first developed in the study of organic chemistry, where the principle is much more conspicuous than among inorganic compounds. Thus, as has been shown (40), we are acquainted with twenty acids resembling acetic acid, which form a series beginning with formic acid and ending with melissic acid. Each member of this series differs in composition from the preceding member by CH_2, or by some multiple of this symbol; and the properties of the compounds vary regularly between the extreme limits, according to well-established laws. Moreover, many other similar, although more limited, series of compounds are known, and the principle realized in these organic series seems to be the true idea of all chemical classification. But, in attempting to apply it to the chemical elements, we find only two or three groups of atoms where the series is of sufficient extent to make the relations of the members evident. In most cases it would seem as if we only knew one or two members of a series, and this apparent ignorance not only throws doubt on the general application of our principle, but also renders uncertain the details of our scheme, even assuming that the principle of the classification is correct. Hence, also, great differences of opinion may be reasonably entertained in regard to the position which the different atoms ought to occupy in such a scheme.

Another very important cause of uncertainty in any scheme of classifying the elements arises from the double relationships which many of them manifest. Thus iron, which we have associated with manganese and aluminum, is in some of its relations closely allied to magnesium and zinc. Many other elements resemble iron in having a similar two-fold character, and different authors may reasonably assign to such elements different places in their systems of classification, according as they chiefly view them from one or the other aspect. Hence arises a degree of uncertainty which affects our whole system, and cannot be avoided in the present state of the science.

Indeed, no classification in independent groups can satisfy the complex relations of the elements. These relations cannot be represented by a simple system of parallel series, but only by a web of crossing lines, in which the same element may be represented as a member of two or more series at once, and as affiliating in different directions with very different classes of elements. In the present fragmentary state of our knowledge, such a classification as we have just indicated is not attainable. The scheme adopted in this book only indicates in each case a single line of relationship; but we have always endeavored to place each element in that relation which is the most characteristic; and, however imperfect such a scheme may be, it will nevertheless assist study by bringing before the student's mind the facts of the science in a systematic and natural order.

104. *Relations of the Atomic Weights.* — If the principle of classification which we have adopted is correct, and the elements actually belong to series like those of the compounds of organic chemistry, we should naturally expect that the atomic weights would conform to the same serial law; and it is a remarkable fact that the differences between the atomic weights of the elements of the same group are in most cases very nearly multiples of 16. The value of this common difference varies between 15 and 17, and we must admit in some cases the simplest fractional multiples; but the mean value is very nearly 16, and the frequent occurrence of this difference is very striking. This numerical relation is not absolutely exact, but here, as in the periods of the planets, in the distribution of leaves on the stem of a plant, and in other similar natural phenomena, there is a marked tendency towards a certain numerical result, which is fully realized, however, only in comparatively few cases.

Other numerical relations which have been noticed between the atomic weights are probably only phases of the same law of distribution in series. Thus the atomic weight of sodium is very nearly the mean between that of lithium and potassium; and the atomic weights of chlorine, bromine, and iodine, of glucinum, yttrium and erbium, of calcium, strontium, and barium, of oxygen, sulphur, and selenium, are similarly related. Again, there are several pairs of allied elements, between whose atomic weights there is very nearly the same difference. Thus the difference between the atomic weights of indium and cadmium is very nearly the same as that between the atomic weights of magnesium and zinc, and the difference between the atomic weights of niobium and tantalum the same as that between the atomic weights of molybdenum and tungsten. A careful study of the atomic weights will also reveal many other approximate relations of the same sort; but although the study of these relations is highly interesting, and may lead hereafter to valuable results, yet no great importance can be attached to them in the present state of the science.

TABLE I.

FRENCH MEASURES.

Measures of Length.

1 Kilometre	= 1000 Metres.	1 Metre	=	1.000 Metre.
1 Hectometre	= 100 "	1 Decimetre	=	0.100 "
1 Decametre	= 10 "	1 Centimetre	=	0.010 "
1 Metre	= 1 "	1 Millimetre	=	0.001 "

		Logarithms.	Ar. Co. Log.
1 Kilometre	= 0.6214 Mile.	9.7933 712	0.2066 188
1 Metre	= 3.2809 Feet.	0.5159 930	9.4840 070
1 Centimetre	= 0.3937 Inch.	9.5951 742	0.4048 258

The metre is one ten-millionth of a quadrant of the globe.

Measures of Volume.

1 Cubic Metre	$\overline{m.}^3$	=	1000.000 Litres.
1 Cubic Decimetre	$\overline{d.m.}^3$	=	1.000 "
1 Cubic Centimetre	$\overline{c.m.}^3$	=	0.001 "

		Logarithms.	Ar. Co. Log.
1 Cubic Metre	= 35.31660 Cubic Feet.	1.5479 790	8.4520 210
1 Cubic Decimetre	= 61.02709 Cubic Inches.	1.7855 226	8.2144 774
1 Cubic Centimetre	= 0.06103 " "	8.7855 226	1.2144 774
1 Litre	= 0.22017 Gallon.	9.3427 581	0.6572 419
1 Litre	= 0.88066 Quart.	9.9448 083	0.0551 917
1 Litre	= 1.76133 Pints.	0.2458 407	9.7541 593

FRENCH WEIGHTS.

1 Kilogramme	= 1000 Grammes.	1 Gramme	=	1.000 Gramme.
1 Hectogramme	= 100 "	1 Decigramme	=	0.100 "
1 Decagramme	= 10 "	1 Centigramme	=	0.010 "
1 Gramme	= 1 "	1 Milligramme	=	0.001 "

		Logarithms.	Ar. Co. Log.
1 Kilogramme	= 2.20462 Pounds Avoirdupois.	0.3433 337	9.6566 663
1 "	= 2.67922 " Troy.	0.4280 083	9.5719 917
1 Gramme	= 15.43235 Grains.	1.1884 321	8.8115 679
1 Crith	= 0.089578 Grammes.	8.9522 014	1.0477 986

TABLE II.

ELEMENTARY ATOMS.

Perissad Elements.	Atomic Weights.	Symbols of Molecules.	Quantivalence.	Artiad Elements.	Atomic Weights.	Symbols of Molecules.	Quantivalence.
Hydrogen	1.0	H-H	I	Copper	63.4	Cu?	II
Fluorine	19.0	F-F	"	Mercury	200.0	Hg	"
Chlorine	35.5	Cl-Cl	"	Calcium	40.0	Ca?	"
Bromine	80.0	Br-Br	"	Strontium	87.6	Sr?	"
Iodine	127.0	I-I	"	Barium	137.0	Ba?	"
Lithium	7.0	Li-Li	"	Lead	207.0	Pb?	"
Sodium	23.0	Na-Na	"	Magnesium	24.0	Mg?	"
Potassium	39.1	K-K	"	Zinc	65.2	Zn?	"
Rubidium	85.4	Rb-Rb	"	Indium	72.0	In?	"
Cæsium	133.0	Cs-Cs	"	Cadmium	112.0	Cd?	"
Silver	108.0	Ag-Ag?	"	Glucinum	9.3	G?	"
Thallium	204.0	Tl-Tl?	I or III	Yttrium	61.7	Y?	"
Gold	197.0	Au=Au?	III	Erbium	112.6	E?	"
Boron	11.0	B=B?	"	Cerium	92.0	Ce?	"
Nitrogen	14.0	N=N	III or V	Lanthanum	93.6	La?	"
Phosphorus	31.0	P=P_2	"	Didymium	95.0	D?	"
Arsenic	75.0	As_2=As_2	"	Nickel	58.8	Ni?	"
Antimony	122.0	Sb_2=Sb_2?	"	Cobalt	58.8	Co?	"
Bismuth	210.0	Bi_2=Bi_2?	"	Manganese	55.0	Mn?	II or IV
Vanadium	51.37	V=V?	"	Iron	56.0	Fe?	"
Uranium	120.0	U=U?	"	Aluminum	27.4	Al?	"
Columbium	94.0	Cb=Cb?	V	Chromium	52.2	Cr?	"
Tantalum	182.0	Ta=Ta?	"	Ruthenium	104.4	Ru?	"
				Osmium	199.2	Os?	"
				Rhodium	104.4	R?	"
				Iridium	196.0	Ir?	"
Artiad Elements.				Palladium	106.6	Pd?	"
				Platinum	197.4	Pt?	"
				Titanium	50.0	Ti?	"
Oxygen	16.0	O=O	II	Tin	118.0	Sn?	"
Sulphur	32.0	S=S	II or VI	Zirconium	89.6	Zr?	IV
Selenium	79.4	Se=Se	"	Thorium	231.4	Th?	"
Tellurium	128.0	Te=Te	"	Silicon	28.0	Si?	"
Molybdenum	96.0	Mo?	VI	Carbon	12.0	C?	"
Tungsten	184.0	W?	"				

TABLE III.

Specific Gravity of Gases and Vapors.

Names.	Symbols.	Sp.Gr. Air = 1.	Sp.Gr. H-H=1.	Half Molecular Weight.	Logarithms.
Air		1.000	14.43		1.1593
Hydrogen	$H-H$	0.0693	1.00	1.00	0.0000
Acetylic Hydride (Aldehyde)	C_2H_3O-H	1.532	22.10	22.00	1.3424
Acetylic Chloride	C_2H_3O-Cl	2.87	41.42	39.25	1.5938
Acetic Anhydride	$(C_2H_3O)_2=O$	3.47	50.07	51.00	1.7076
Acetic Acid	$H-O-C_2H_3O$	2.083	30.07	30.00	1.4771
Aluminic Chloride	$[Al_2] \equiv Cl_6$	9.34	134.80	133.90	2.1268
Aluminic Bromide	$[Al_2] \equiv Br_6$	18.62	268.70	267.40	2.4272
Aluminic Iodide	$[Al_2] \equiv I_6$	27.	389.60	408.40	2.6111
Antimonious Chloride	$Sb \equiv Cl_3$	7.8	112.70	114.20	2.0577
Triethylstibine	$(C_2H_5)_3 \equiv Sb$	7.23	104.40	104.50	2.0191
Arsenic	$As_2 \equiv As_2$	10.6	153.00	150.00	2.1761
Arseniuretted Hydrogen	$H_3 \equiv As$	2.695	38.90	39.00	1.5911
Triethylarsine	$(C_2H_5)_3 \equiv As$	5.29	76.35	81.00	1.9085
Kakodyl	$(CH_3)_2As-(CH_3)_2As$	7.10	102.50	105.00	2.0212
Arsenious Chloride	$As \equiv Cl_3$	6.3	90.90	90.75	1.9578
Arsenious Iodide	$As \equiv I_3$	16.1	232.40	228.00	2.3579
Bismuthous Chloride	$Bi \equiv Cl_3$	11.35	163.90	158.25	2.1994
Boric Methide	$(CH_3)_3 \equiv B$	1.931	27.90	28.00	1.4472
Boric Ethide	$(C_2H_5)_3 \equiv B$	3.401	49.10	49.00	1.6902
Boric Fluoride	$B \equiv F_3$	2.37	34.20	34.00	1.5315
Boric Chloride	$B \equiv Cl_3$	3.942	56.85	58.75	1.7690
Boric Bromide	$B \equiv Br_3$	8.78	126.80	125.50	2.0986
Methylic Borate	$(CH_3)_3 \equiv O_3 \equiv B$	3.59	51.80	52.00	1.7160
Ethylic Borate	$(C_2H_5)_3 \equiv O_3 \equiv B$	5.14	74.20	73.00	1.8633
Bromine	$Br-Br$	5.54	79.50	80.00	1.9031
Hydrobromic Acid	$H-Br$	2.71	39.10	40.50	1.6075
Carbonic Tetrachloride	$C \equiv Cl_4$	5.415	78.14	77.00	1.8865
Carbonic Oxydichloride (Phosgene Gas)	$C \equiv O, Cl_2$	3.399	49.06	49.50	1.6946
Dicarbonic Hexachloride	$[C-C] \equiv Cl_6$	8.157	117.70	118.50	2.0737
Dicarbonic Tetrachloride	$[C=C] \equiv Cl_4$	5.82	84.00	83.00	1.9191
Dicarbonic Dichloride	$[C \equiv C] = Cl_2$			47.50	1.6767
Carbonic Oxide	$C=O$	0.967	13.95	14.00	1.1461
Carbonic Anhydride	$C \equiv O_2$	1.529	22.06	22.00	1.3424
Carbonic Sulphide	$C \equiv S_2$	2.645	38.17	38.00	1.5798
Chlorine	$Cl-Cl$	2.44	35.22	35.50	1.5502
Hydrochloric Acid	$H-Cl$	1.27	18.32	18.25	1.2613
Chromic Oxychloride	$[Cr_2] \equiv O_2, Cl_2$	5.5	79.40	78.25	1.8935
Columbic Chloride	$Cb \equiv Cl_5$	9.6	138.60	135.70	2.1326
Columbic Oxychloride	$Cb \equiv O, Cl_3$	7.9	114.00	108.20	2.0342
Cyanogen	$CN-CN$	1.806	26.06	26.00	1.4150
Hydrocyanic Acid	$H-CN$	0.947	13.67	13.50	1.1303
Ethyl	$C_2H_5-C_2H_5$	2.0	28.86	29.00	1.4624
Ethylic Chloride	$(C_2H_5)-Cl$	2.219	32.02	32.25	1.5085
Ethylic Oxide (Ether)	$(C_2H_5)_2=O$	2.586	37.32	37.00	1.5682
Ethylic Hydrate (Alcohol)	C_2H_5-O-H	1.613	23.28	23.00	1.3617

TABLE III. (*Continued.*)

Names.	Symbols.	Sp.Gr. Air =1.	Sp.Gr. H-H=1.	Half Molecular Weight.	Logarithms.
Ethylene (Olefiant Gas)	C_2H_4	0.978	14.11	14.00	1.1461
" Chloride (Dutch Liq.)	$(C_2H_4)=Cl_2$	3.443	49.69	49.50	1.6946
Ethylene Oxide	$(C_2H_4)=O$	1.422	20.52	22.00	1.3424
Ethylene Hydrate (Glycol)	$(C_2H_4)=O_2=H_2$			31.00	1.4914
Ferric Chloride	$[Fe_2]\equiv Cl_6$	11.39	164.40	162.50	2.2108
Iodine	$I-I$	8.716	125.90	127.00	2.1038
Hydriodic Acid	$H-I$	4.443	64.12	64.00	1.8062
Mercury	Hg	6.976	100.70	100.00	2.0000
Mercuric Ethide	$(C_2H_5)_2=Hg$	9.97	143.90	129.00	2.1106
Mercuric Methide	$(CH_3)_2=Hg$	8.29	119.60	115.00	2.0607
Mercuric Chloride	$Hg=Cl_2$	9.8	141.50	135.50	2.1319
Mercuric Bromide	$Hg=Br_2$	12.16	175.60	180.00	2.2553
Mercuric Iodide	$Hg=I_2$	15.9	229.60	227.00	2.3560
Mercurous Chloride	$[Hg_2]=Cl_2$	8.21	118.50	235.50	2.3720
Nitrogen	$N\equiv N$	0.971	14.00	14.00	1.1461
Ammonia	$H_3\equiv N$	0.591	8.535	8.51	0.9294
Methylamine	$H_2,(CH_3)\equiv N$	1.08	15.59	15.50	1.1903
Aniline	$H_2,(C_6H_5)\equiv N$	3.21	46.83	46.50	1.6675
Nitrous Oxide	N_2O	1.527	22.04	22.00	1.3424
Nitric Oxide	NO	1.038	14.97	15.00	1.1761
Nitric Peroxide	NO_2	1.72	24.82	23.00	1.3617
Osmic Tetroxide	OsO_4	8.89	128.30	131.60	2.1193
Oxygen	$O=O$	1.1056	15.95	16.00	1.2041
Aqueous Vapor	$H_2=O$	0.6235	8.998	9.00	0.9542
Phosphorus	$P_2\equiv P_2$	4.42	63.78	62.00	1.7924
Phosphuretted Hydrogen	$H_3\equiv P$	1.184	17.09	17.00	1.2304
Phosphorous Chloride	$P\equiv Cl_3$	4.742	68.44	68.75	1.8373
Phosphoric Oxychloride	$P\equiv O, Cl_3$	5.3	76.49	76.75	1.8851
Oxide of Triethylphosphine	$((C_2H_5)_3\equiv P)=O$	4.6	66.39	67.00	1.8261
Selenium, at 771°	$Se=Se$	5.68	81.96	79.40	1.8998
Seleniuretted Hydrogen	$H_2=Se$	2.795	40.33	40.70	1.6096
Silicic Methide	$(CH_3)_4\equiv Si$	3.083	44.49	44.00	1.6435
Silicic Ethide	$(C_2H_5)_4\equiv Si$	5.13	74.03	72.00	1.8573
Silicic Fluoride	$Si\equiv F_4$	3.600	51.95	52.00	1.7160
Silicic Chloride	$Si\equiv Cl_4$	5.939	85.72	85.00	1.9294
Ethylic Silicate	$(C_2H_5)_4\equiv O_4\equiv Si$	7.32	105.60	104.00	1.0170
Stannic Ethide	$(C_2H_5)_4\equiv Sn$	8.021	115.80	117.00	2.0682
Stannic Dimethylo-diethide	$(CH_3)_2,(C_2H_5)_2\equiv Sn$	6.838	98.68	103.00	2.0128
Stannic Chloro-triethide	$Cl,(C_2H_5)_3\equiv Sn$	8.430	121.70	120.20	2.0799
Stannic Dichloro-diethide	$Cl_2,(C_2H_5)_2\equiv Sn$	8.710	125.70	123.50	2.0917
Stannic Chloride	$Sn\equiv Cl_4$	9.199	132.70	130.00	2.1139
Sulphur above 860°	$S=S$	2.23	32.18	32.00	1.5051
Sulphur at 450°	S_6	6.617	95.50	96.00	1.9823
Sulphuretted Hydrogen	$H_2=S$	1.191	17.19	17.00	1.2304
Sulphurous Anhydride	$S\equiv O_2$	2.234	32.24	32.00	1.5051
Sulphuric Anhydride	$S\equiv O_3$	2.763	39.87	40.00	1.6021
Tantalic Chloride	$TaCl_5$	12.8	184.70	179.70	2.2546
Titanic Chloride	$TiCl_4$	6.836	98.65	96.00	1.9823
Zinc Ethide	$(C_2H_5)_2=Zn$	4.259	61.46	61.60	1.7896
Zirconic Chloride	$Zr\equiv Cl_4$	8.15	117.60	115.80	2.0637

LOGARITHMS AND ANTILOGARITHMS.

LOGARITHMS OF NUMBERS.

Natural Numbers	0	1	2	3	4	5	6	7	8	9	Proportional Parts.								
											1	2	3	4	5	6	7	8	9
10	0000	0043	0086	0128	0170	0212	0253	0294	0334	0374	4	8	12	17	21	25	29	33	37
11	0414	0453	0492	0531	0569	0607	0645	0682	0719	0755	4	8	11	15	19	23	26	30	34
12	0792	0828	0864	0899	0934	0969	1004	1038	1072	1106	3	7	10	14	17	21	24	28	31
13	1139	1173	1206	1239	1271	1303	1335	1367	1399	1430	3	6	10	13	16	19	23	26	29
14	1461	1492	1523	1553	1584	1614	1644	1673	1703	1732	3	6	9	12	15	18	21	24	27
15	1761	1790	1818	1847	1875	1903	1931	1959	1987	2014	3	6	8	11	14	17	20	22	25
16	2041	2068	2095	2122	2148	2175	2201	2227	2253	2279	3	5	8	11	13	16	18	21	24
17	2304	2330	2355	2380	2405	2430	2455	2480	2504	2529	2	5	7	10	12	15	17	20	22
18	2553	2577	2601	2625	2648	2672	2695	2718	2742	2765	2	5	7	9	12	14	16	19	21
19	2788	2810	2833	2856	2878	2900	2923	2945	2967	2989	2	4	7	9	11	13	16	18	20
20	3010	3032	3054	3075	3096	3118	3139	3160	3181	3201	2	4	6	8	11	13	15	17	19
21	3222	3243	3263	3284	3304	3324	3345	3365	3385	3404	2	4	6	8	10	12	14	16	18
22	3424	3444	3464	3483	3502	3522	3541	3560	3579	3598	2	4	6	8	10	12	14	15	17
23	3617	3636	3655	3674	3692	3711	3729	3747	3766	3784	2	4	6	7	9	11	13	15	17
24	3802	3820	3838	3856	3874	3892	3909	3927	3945	3962	2	4	5	7	9	11	12	14	16
25	3979	3997	4014	4031	4048	4065	4082	4099	4116	4133	2	3	5	7	9	10	12	14	15
26	4150	4166	4183	4200	4216	4232	4249	4265	4281	4298	2	3	5	7	8	10	11	13	15
27	4314	4330	4346	4362	4378	4393	4409	4425	4440	4456	2	3	5	6	8	9	11	13	14
28	4472	4487	4502	4518	4533	4548	4564	4579	4594	4609	2	3	5	6	8	9	11	12	14
29	4624	4639	4654	4669	4683	4698	4713	4728	4742	4757	1	3	4	6	7	9	10	12	13
30	4771	4786	4800	4814	4829	4843	4857	4871	4886	4900	1	3	4	6	7	9	10	11	13
31	4914	4928	4942	4955	4969	4983	4997	5011	5024	5038	1	3	4	6	7	8	10	11	12
32	5051	5065	5079	5092	5105	5119	5132	5145	5159	5172	1	3	4	5	7	8	9	11	12
33	5185	5198	5211	5224	5237	5250	5263	5276	5289	5302	1	3	4	5	6	8	9	10	12
34	5315	5328	5340	5353	5366	5378	5391	5403	5416	5428	1	3	4	5	6	8	9	10	11
35	5441	5453	5465	5478	5490	5502	5514	5527	5539	5551	1	2	4	5	6	7	9	10	11
36	5563	5575	5587	5599	5611	5623	5635	5647	5658	5670	1	2	4	5	6	7	8	10	11
37	5682	5694	5705	5717	5729	5740	5752	5763	5775	5786	1	2	3	5	6	7	8	9	10
38	5798	5809	5821	5832	5843	5855	5866	5877	5888	5899	1	2	3	5	6	7	8	9	10
39	5911	5922	5933	5944	5955	5966	5977	5988	5999	6010	1	2	3	4	5	7	8	9	10
40	6021	6031	6042	6053	6064	6075	6085	6096	6107	6117	1	2	3	4	5	6	8	9	10
41	6128	6138	6149	6160	6170	6180	6191	1201	6212	6222	1	2	3	4	5	6	7	8	9
42	6232	6243	6253	6263	6274	6284	6294	6304	9314	6325	1	2	3	4	5	6	7	8	9
43	6335	6345	6355	6365	6375	6385	6395	6405	6415	6425	1	2	3	4	5	6	7	8	9
44	6435	6444	6454	6464	6474	6484	6493	6503	6513	6522	1	2	3	4	5	6	7	8	9
45	6532	6542	6551	6561	6571	6580	6590	6599	6609	6618	1	2	3	4	5	6	7	8	9
46	6628	6637	6646	6656	6665	6675	6684	6693	6702	6712	1	2	3	4	5	6	7	7	8
47	6721	6730	6739	6749	6758	6767	6776	6785	6794	6803	1	2	3	4	5	5	6	7	8
48	6812	6821	6830	6839	6848	6857	6866	6875	6884	6893	1	2	3	4	4	5	6	7	8
49	6902	6911	6920	6928	6937	6946	6955	6964	6972	6981	1	2	3	4	4	5	6	7	8
50	6990	6998	7007	7016	7024	7033	7042	7050	7059	7067	1	2	3	3	4	5	6	7	8
51	7076	7084	7093	7101	7110	7118	7126	7135	7143	7152	1	2	3	3	4	5	6	7	8
52	7160	7168	3177	7185	7193	7202	7210	7218	7226	7235	1	2	2	3	4	5	6	7	7
53	7243	7251	7259	7267	7275	7284	7292	7300	7308	7316	1	2	2	3	4	5	6	6	7
54	7324	7332	7340	7348	7356	7364	7372	7380	7388	7396	1	2	2	3	4	5	6	6	7

LOGARITHMS OF NUMBERS.

Natural Numbers	0	1	2	3	4	5	6	7	8	9	Proportional Parts.								
											1	2	3	4	5	6	7	8	9
55	7404	7412	7419	7427	7435	7443	7451	7459	7466	7474	1	2	2	3	4	5	5	6	7
56	7482	7490	7497	7505	7513	7520	7528	7536	7543	7551	1	2	2	3	4	5	5	6	7
57	7559	7566	7574	7582	7589	7597	7604	7612	7619	7627	1	2	2	3	4	5	5	6	7
58	7634	7642	7649	7657	7664	7672	7679	7686	7694	7701	1	1	2	3	4	4	5	6	7
59	7709	7716	7723	7731	6738	7745	7752	7760	7767	7774	1	1	2	3	4	4	5	6	7
60	7782	7789	7796	7803	7810	7818	7825	7832	7839	7846	1	1	2	3	4	4	5	6	6
61	7853	7860	7868	7875	7882	7889	7896	7903	7910	7917	1	1	2	3	4	4	5	6	6
62	7924	7931	7938	7945	7952	7959	7966	7973	7980	8987	1	1	2	3	3	4	5	6	6
63	7993	8000	8007	8014	8021	8028	8035	8041	8048	8055	1	1	2	3	3	4	5	5	6
64	8062	8069	8075	8082	8089	8096	8102	8109	8116	8122	1	1	2	3	3	4	5	5	6
65	8129	8136	8142	8149	8156	8162	8169	8176	8182	8189	1	1	2	3	3	4	5	5	6
66	8195	8202	8209	8215	8222	8228	8235	8241	8248	8254	1	1	2	3	3	4	5	5	6
67	8261	8267	8274	8280	8287	8293	8299	8306	8312	8319	1	1	2	3	3	4	5	5	6
68	8325	8331	8338	8344	8351	8357	8363	9370	8376	8382	1	1	2	3	3	4	4	5	6
69	8388	8395	8401	8407	8414	8420	8426	8432	8439	8445	1	1	2	2	3	4	4	5	6
70	8451	8457	8463	8470	8476	8482	8488	8494	8500	8506	1	1	2	2	3	4	4	5	6
71	8513	8519	8525	8531	8537	8543	8549	8555	8561	8567	1	1	2	2	3	4	4	5	5
72	8573	8579	8585	8591	8597	8603	8609	8615	8621	8627	1	1	2	2	3	4	4	5	5
73	8633	8639	8645	8651	8657	8663	8669	8675	8681	8696	1	1	2	2	3	4	4	5	5
74	8692	8698	8704	8710	8716	8722	8727	8733	8739	8745	1	1	2	2	3	4	4	5	5
75	8751	8756	8762	8768	8774	8779	8785	8791	8797	8802	1	1	2	2	3	3	4	5	5
76	8808	8814	8820	8825	8831	8837	8842	8848	8854	8859	1	1	2	2	3	3	4	5	5
77	8865	8871	8876	8882	8887	8893	8899	8904	8910	8915	1	1	2	2	3	3	4	4	5
78	8921	8927	8932	8938	8943	8949	8954	8960	8965	8971	1	1	2	2	3	3	4	4	5
79	8976	8982	8987	8993	8998	9004	9009	9015	9020	9025	1	1	2	2	3	3	4	4	5
80	9031	9036	6042	9047	9053	9058	9063	9069	9074	9079	1	1	2	2	3	3	4	4	5
81	9085	9090	9096	9101	9106	9112	9117	9122	9128	9133	1	1	2	2	3	3	4	4	5
82	9138	9143	9149	9154	9159	9165	9170	9175	9180	9186	1	1	2	2	3	3	4	4	5
83	9191	9196	9201	9206	9212	9217	9222	9227	9232	9238	1	1	2	2	3	3	4	4	5
84	9243	9248	9253	9258	9263	9269	9274	9279	9384	9289	1	1	2	2	3	3	4	4	5
85	9294	9299	9304	9309	9315	9320	9325	9330	9335	9340	1	1	2	2	3	3	4	4	5
86	9345	9350	9355	9360	9465	9370	9375	9380	9385	9390	1	1	2	2	3	3	4	4	5
87	9395	9400	9405	9410	9415	9420	9325	9430	9435	9440	0	1	1	2	2	3	3	4	4
88	9445	9450	9455	9460	9465	9469	9474	9479	9484	9489	0	1	1	2	2	3	3	4	4
89	9494	9499	9504	9509	9513	9518	9523	9528	9533	9538	0	1	1	2	2	3	3	4	4
90	9542	9547	9552	9557	9562	9566	9571	9576	9581	9586	0	1	1	2	2	3	3	4	4
91	9590	9595	9600	9605	9609	9614	9619	9624	9628	9633	0	1	1	2	2	3	3	4	4
92	9638	9843	9647	9652	9657	9661	9666	9671	9675	9680	0	1	1	2	2	3	3	4	4
93	9685	9689	9694	9699	9703	9708	9713	9717	9722	9727	0	1	1	2	2	3	3	4	4
94	9731	9736	9741	9745	9750	9754	9759	9763	9768	9773	0	1	1	2	2	3	3	4	4
95	9777	9782	9786	9791	9795	9800	9805	9809	9814	9818	0	1	1	2	2	3	3	4	4
96	9823	9827	9832	9836	9841	9845	9850	9854	9859	9863	0	1	1	2	2	3	3	4	4
97	9868	9872	9877	9881	9886	9890	9894	9899	9903	9908	0	1	1	2	2	3	3	4	4
98	9912	9917	9921	9926	9930	9934	9939	9943	9948	9952	0	1	1	2	2	3	3	4	4
99	9956	9961	9965	9969	9974	9978	9983	9987	9991	9996	0	1	1	2	2	3	3	3	4

ANTILOGARITHMS.

Logarithm	0	1	2	3	4	5	6	7	8	9	1	2	3	4	5	6	7	8	9
.00	1000	1002	1005	1007	1009	1012	1014	1016	1019	1021	0	0	1	1	1	1	2	2	2
.01	1023	1026	1028	1030	1033	1035	1038	1040	1042	1045	0	0	1	1	1	1	2	2	2
.02	1047	1050	1052	1054	1057	1059	1062	1064	1067	1069	0	0	1	1	1	1	2	2	2
.03	1072	1074	1076	1079	1081	1084	1086	1089	1091	1094	0	0	1	1	1	1	2	2	2
.04	1096	1099	1102	1104	1107	1109	1112	1114	1117	1119	0	1	1	1	1	2	2	2	2
.05	1122	1125	1127	1130	1132	1135	1138	1140	1143	1146	0	1	1	1	1	2	2	2	2
.06	1148	1151	1153	1156	1159	1161	1164	1167	1169	1172	0	1	1	1	1	2	2	2	2
.07	1175	1178	1180	1183	1186	1189	1191	1194	1197	1199	0	1	1	1	1	2	2	2	2
.08	1202	1205	1208	1211	1213	1216	1219	1222	1225	1227	0	1	1	1	1	2	2	2	3
.09	1230	1233	1236	1239	1242	1245	1247	1250	1253	1256	0	1	1	1	1	2	2	2	3
.10	1259	1262	1265	1268	1271	1274	1276	1579	1282	1285	0	1	1	1	1	2	2	2	3
.11	1288	1291	1294	1297	1300	1303	1306	1309	1312	1215	0	1	1	1	2	2	2	2	3
.12	1318	1321	1324	1327	1330	1334	1337	1340	1343	1346	0	1	1	1	2	2	2	2	3
.13	1349	1352	1355	1358	1361	1365	1368	1371	1374	1377	0	1	1	1	2	2	2	3	3
.14	1380	1384	1387	1390	1393	1396	1400	1403	1406	1409	0	1	1	1	2	2	2	3	3
.15	1413	1416	1419	1422	1426	1429	1432	1435	1439	1442	0	1	1	1	2	2	2	3	3
.16	1445	1449	1452	1455	1459	1462	1466	1469	1472	1476	0	1	1	1	2	2	2	3	3
.17	1479	1483	1486	1489	1493	1496	1500	1503	1507	1510	0	1	1	1	2	2	2	3	3
.18	1514	1517	1521	1524	1528	1531	1535	1538	1542	1545	0	1	1	1	2	2	2	3	3
.19	1549	1552	1556	1560	1563	1567	1570	1574	1578	1581	0	1	1	1	2	2	3	3	3
.20	1585	1589	1592	1596	1600	1603	1607	1611	1614	1618	0	1	1	1	2	2	3	3	3
.21	1622	1626	1629	1633	1637	1641	1644	1648	1652	1656	0	1	1	2	2	2	3	3	3
.22	1660	1663	1667	1671	1675	1679	1683	1687	1690	1694	0	1	1	2	2	2	3	3	3
.23	1698	1702	1706	1710	1714	1718	1722	1726	1730	1734	0	1	1	2	2	2	3	3	4
.24	1738	1742	1746	1750	1754	1758	1762	1766	1770	1774	0	1	1	2	2	2	3	3	4
.25	1778	1782	1786	1791	1795	1799	1803	1807	1811	1816	0	1	1	2	2	2	3	3	4
.26	1820	1324	1828	1832	1837	1841	1845	1849	1854	1858	0	1	1	2	2	3	3	3	4
.27	1862	1866	1871	1875	1879	1884	1888	1892	1897	1901	0	1	1	2	2	3	3	3	4
.28	1905	1910	1914	1919	1923	1928	1932	1936	1341	1945	0	1	1	2	2	3	3	4	4
.29	1950	1954	1959	1963	1968	1972	1977	1982	1986	1991	0	1	1	2	2	3	3	4	4
.30	1995	2000	2004	2009	2014	2018	2023	2038	1032	2037	0	1	1	2	2	3	3	4	4
.31	2042	2046	2051	2056	2061	2065	2070	2075	2080	2084	0	1	1	2	2	3	3	4	4
.32	2089	2094	2099	2104	2109	2113	2118	2123	2128	2133	0	1	1	2	2	3	3	4	4
.33	2138	2143	2148	2153	2158	2163	2168	2173	2178	2183	0	1	1	2	2	3	3	4	4
.34	2188	2193	2198	2203	2208	2213	2218	2223	2228	2234	1	1	2	2	3	3	4	4	5
.35	2239	2244	2249	2254	2259	2265	2270	2275	2280	2296	1	1	2	2	3	3	4	4	5
.36	2291	2296	2301	2307	2312	2317	2323	2328	2333	2339	1	1	2	2	3	3	4	4	5
.37	2344	2350	2355	2360	2366	2371	2677	2382	2388	2393	1	1	2	2	3	3	4	4	5
.38	2399	2404	2410	2415	2421	2427	2432	2438	2443	2449	1	1	2	2	3	3	4	4	5
.39	2455	2460	2466	2472	2477	2483	2489	2495	2500	2506	1	1	2	2	3	3	4	5	5
.40	2512	2518	2523	2529	2535	2541	2547	2553	2559	2564	1	1	2	2	3	4	4	5	5
.41	2570	2576	2582	2588	2594	2600	2606	2612	2618	2624	1	1	2	2	3	4	4	5	5
.42	2630	2636	2642	2649	2655	2661	2667	2673	2679	2685	1	1	2	3	3	4	4	5	6
.43	2692	2698	2704	2710	2716	2723	2729	2735	2742	2748	1	1	2	3	3	4	4	5	6
.44	2754	2761	2767	2773	2780	2786	2793	2799	2805	2812	1	1	2	3	3	4	4	5	6
.45	2818	2825	2831	2838	2844	2851	2858	2864	2871	2877	1	1	2	3	3	4	5	5	6
.46	2884	2891	2897	2904	2911	2917	2924	2931	2938	2944	1	1	2	3	3	4	5	5	6
.47	2951	2958	2965	2972	2979	2985	2992	2999	3006	3013	1	1	2	3	3	4	5	5	6
.48	3020	3027	3034	3041	3048	3055	3062	3069	3076	3083	1	1	2	3	4	4	5	5	6
.49	2090	3097	3105	3112	3119	3126	3133	3141	3148	3155	1	1	2	3	4	4	5	6	6

ANTILOGARITHMS.

Logarithms.	0	1	2	3	4	5	6	7	8	9	Proportional Parts.								
											1	2	3	4	5	6	7	8	9
.50	3162	3170	3177	3184	3192	3199	3206	3214	3221	3228	1	1	2	3	4	4	5	6	7
.51	3236	3243	3251	3258	3266	3273	3281	3289	3296	3304	1	2	2	3	4	5	5	6	7
.52	3311	3319	3327	3334	3342	3350	3357	3365	3373	3381	1	2	2	3	4	5	5	6	7
.53	3388	3396	3404	3412	3420	3428	3436	3443	3451	3459	1	2	2	3	4	5	6	6	7
.54	3467	3475	3483	3491	3499	3508	3516	3524	3532	3540	1	2	2	3	4	5	6	6	7
.55	3548	3556	3565	3573	3581	3589	3597	3606	3614	3622	1	2	2	3	4	5	6	7	7
.56	3631	3639	3648	3656	3664	3673	3681	3690	3698	3707	1	2	3	3	4	5	6	7	8
.57	3715	3724	3733	3741	3750	3758	3767	3776	3784	3793	1	2	3	3	4	5	6	7	8
.58	3802	3811	3819	3828	3837	3846	3855	3864	3873	3882	1	2	3	4	4	5	6	7	8
.59	3890	3899	3908	3917	3926	3936	3945	3954	3963	3972	1	2	3	4	5	5	6	7	8
.60	3981	3990	3999	4009	4018	4027	4036	4046	4055	4064	1	2	3	4	5	6	6	7	8
.61	4074	4083	4093	4102	4111	4121	4130	4140	4150	4159	1	2	3	4	5	6	7	8	9
.62	4169	4178	4188	4198	4207	4217	4227	4236	4246	4256	1	2	3	4	5	6	7	8	9
.63	4266	4276	4285	4295	4305	4315	4325	4335	4345	4355	1	2	3	4	5	6	7	8	9
.64	4365	4375	4385	4395	4406	4416	4426	4436	4446	4457	1	2	3	4	5	6	7	8	9
.65	4467	4477	4487	4498	4508	4519	4529	4539	4550	4560	1	2	3	4	5	6	7	8	9
.66	4571	4581	4592	4603	4613	4624	4634	4645	4656	4667	1	2	3	4	5	6	7	9	10
.67	4677	4688	4699	4710	4721	4732	4742	4753	4764	4775	1	2	3	4	5	7	8	9	10
.68	4786	4797	4808	4819	4831	4842	4853	4864	4875	4887	1	2	3	4	6	7	8	9	10
.69	4898	4909	4920	4932	4943	4955	4966	4977	4989	5000	1	2	3	5	6	7	8	9	10
.70	5012	5023	5035	5047	5058	5070	5082	5093	5105	5117	1	2	4	5	6	7	8	9	11
.71	5129	5140	5152	5164	5176	5188	5200	5212	5224	5236	1	2	4	5	6	7	8	10	11
.72	5248	5260	5272	5284	5297	5309	5321	5333	5346	5358	1	2	4	5	6	7	9	10	11
.73	5370	5383	5395	5408	5420	5433	5445	5458	5470	5483	1	3	4	5	6	8	9	10	11
.74	5495	5508	5521	5534	5546	5559	5572	5585	5598	5610	1	3	4	5	6	8	9	10	12
.75	5623	5636	5649	5662	5675	5689	5702	5715	5728	5741	1	3	4	5	7	8	9	10	12
.76	5754	5768	5781	5794	5808	5821	5834	5848	5861	5875	1	3	4	5	7	8	9	11	12
.77	5888	5902	5916	5929	5943	5957	5970	5984	5998	6012	1	3	4	5	7	8	10	11	12
.78	6026	6039	6053	6067	6081	6095	6109	6124	6138	6152	1	3	4	6	7	8	10	11	13
.79	6166	6180	6194	6209	6223	6237	6252	6266	6281	6295	1	3	4	6	7	9	10	11	13
.80	6310	6324	6339	6353	6368	6383	6397	6412	6427	6442	1	3	4	6	7	9	10	12	13
.81	6457	6471	6486	6501	6516	6531	6546	6561	6577	6592	2	3	5	6	8	9	11	12	14
.82	6607	6622	6637	6653	6668	6683	6699	6714	6730	6745	2	3	5	6	8	9	11	12	14
.83	6761	6776	6792	6808	6823	6839	6855	6871	6887	6902	2	3	5	6	8	9	11	13	14
.84	6918	6934	6950	6966	6982	6998	7015	7031	7047	7063	2	3	5	6	8	10	11	13	15
.85	7079	7096	7112	7129	7145	7161	7178	7194	7211	7228	2	3	5	7	8	10	12	13	15
.86	7244	7261	7278	7295	7311	7328	7345	7362	7379	7396	2	3	5	7	8	10	12	13	15
.87	7413	7430	7447	7464	7482	7499	7516	7534	7551	7568	2	3	5	7	9	10	12	14	16
.88	7586	7603	7621	7638	7656	7674	7691	7709	7727	7745	2	4	5	7	9	11	12	14	16
.89	7762	7780	7798	7816	7834	7852	7870	7889	7907	7925	2	4	5	7	9	11	13	14	16
.90	7943	7962	7980	7998	8017	8035	8054	8072	8091	8110	2	4	6	7	9	11	13	15	17
.91	8128	8147	8166	8185	8204	8222	8241	8260	8279	8299	2	4	6	8	9	11	13	15	17
.92	8318	8337	8356	8375	8395	8414	8433	8453	8472	8492	2	4	6	8	10	12	14	15	17
.93	8511	8531	8551	8570	8590	8610	8630	8650	8670	8690	2	4	6	8	10	12	14	16	18
.94	8710	8730	8750	8770	8790	8810	8831	8851	8872	8892	2	4	6	8	10	12	14	16	18
.95	8913	8933	8954	8974	8995	9016	9036	9057	9078	9099	2	4	6	8	10	12	15	17	19
.96	9120	9141	9162	9183	9204	9226	9247	9268	9290	9311	2	4	6	8	11	13	15	17	19
.97	9333	9354	9376	9397	9419	9441	9462	9484	9506	9528	2	4	7	9	11	13	15	17	20
.98	9550	9572	9594	9616	9638	9661	9683	9705	9727	9750	2	4	7	9	11	13	16	18	20
.99	9772	9795	9817	9840	9863	9886	9908	9931	9954	9977	2	5	7	9	11	14	16	18	20

CONSTANT LOGARITHMS.

		Logarithms.	Ar. Co. Log.
Circumf. of circle when $R = 1$,	$(\frac{\pi}{2} = 1.5708)$	0.1961	9.8039
" " " " $D = 1$,	$(\pi = 3.1416)$	0.4971	9.5028
Area of circle when $R^2 = 1$,	$(\pi = 3.1416)$	0.4971	9.5028
" " " " $D^2 = 1$,	$(\frac{\pi}{4} = 0.7854)$	9.8951	0.1049
" " " " $C^2 = 1$,	$(\frac{1}{4\pi} = 0.0796)$	8.9008	1.0992
Surface of sphere when $R^2 = 1$,	$(4\pi = 12.5664)$	1.0992	8.9008
" " " " $D^2 = 1$,	$(\pi = 3.1416)$	0.4971	9.5028
" " " " $C^2 = 1$,	$(\frac{1}{\pi} = 0.3183)$	9.5028	0.4971
Solidity of sphere when $R^3 = 1$,	$(\frac{4}{3}\pi = 4.1888)$	0.6221	9.3779
" " " " $D^3 = 1$,	$(\frac{\pi}{6} = 0.5236)$	9.7190	0.2810
" " " " $C^3 = 1$,	$(\frac{1}{6\pi^2} = 0.0169)$	8.2275	1.7724
Weight of one litre of Hydrogen	(0.0896 grammes)	8.9522	1.0478
" " " " " Air	(1.293 ")	0.1116	9.8884
" " " " " "	(14.43 criths)	1.1594	8.8406
Per cent of Oxygen in air by weight	(0.2318)	9.3651	0.6349
" " " Nitrogen " " " "	(0.7682)	9.8855	0.1145
Mean height of Barometer	(76 c. m.)	1.8808	8.1192
Coefficient of expansion of Air	(0.00366)	7.5635	2.4365
Latent Heat of Water	(79)	1.8976	8.1024
" " " Free Steam	(537)	2.7300	7.2700
To reduce 𝔖𝔭.𝔊𝔯. to Sp. Gr., or reverse, add to log.		1.1594 or	8.8406
" " Sp. Gr. to *Sp. Gr.*, " " " " "		6.9522 or	3.0478
" " 𝔖𝔭.𝔊𝔯. to *Sp. Gr.*, " " " " "		7.1116 or	2.8884
" " grammes to criths, " " " " "		1.0478 or	8.9522

www.ingramcontent.com/pod-product-compliance
Lightning Source LLC
Chambersburg PA
CBHW031814220426
43662CB00007B/649